Plumbing

BY RICHARD DEMSKE

PUBLISHERS • GROSSET & DUNLAP • NEW YORK

Acknowledgments

Cover photograph by Mort Engel

For their valued cooperation in the preparation of this book, we wish to express our appreciation to these firms:

American Olean Tile Co.; American-Standard; Eljer Plumbingware; Fluidmaster Inc.; Formco Inc.; General Electric; Kohler Co.; Loctite Corp.; Marlite Paneling; Mobile International, Inc.; Nevamar; Peerless Faucet, Div. of Masco Corp.; Price Pfister; Rockwell International Building Components Division; Rubbermaid, Inc.; Sears, Roebuck and Co.; Stanadyne; Universal-Rundle Corp.

1976 PRINTING

Published simultaneously in Canada
Library of Congress catalog card number: 74-33762
ISBN: 0-448-11956-0 (trade edition)
ISBN: 0-448-13302-4 (library edition)

Printed in the United States of America

Contents

1
The Household Plumbing System

The same home handyman who considers himself perfectly proficient with hammer, saw, and 2 × 4s and a master at backyard bricklaying projects often cringes in terror when faced with a leaking faucet or a clogged and overflowing toilet. There is a certain mystique to plumbing, and many do-it-yourselfers call for the plumber at the first sign of a drip. They pay dearly for their timidity when the plumber arrives and presents his bill for $18.05 ($ .05 for a faucet washer, $18 for labor).

That is, *when* he arrives. Plumbers still do make house calls, but their services are much in demand and it may take a while before he gets around to your problem. That drip may develop into a deluge by the time he comes to the rescue, and the overflowing toilet may float your house off its foundation if the plumber can't make it until "a week from next Thursday." Of course, you can always turn off all the water in the house while you are waiting, but that is likely to cause some family discomfort, to say the least.

There is a better solution: do your own simple plumbing repairs—and more ambitious projects, too.

How Your Household Plumbing System Works

Part of the mystery of plumbing is that so much of it is hidden out of sight, identifiable only by a slight gurgle heard through the walls when somebody draws a bath or flushes a toilet in an upstairs bathroom. While most of the repairs that you make will be on visible components of the plumbing system—faucets, toilets, exposed pipes in the basement—it is helpful to understand how the entire system works. This information is especially important if you plan to make major improvements; for example, installing a new washing machine or adding a bathroom.

The household plumbing system consists of three parts: the supply (where the water comes from), the fixtures (where the water is used), and the drain-waste-vent (where the water goes after it is used).

Convenience and beauty are the hallmarks of the modern bathroom. Far less glamorous—but much more important—is the complex system of supply and drainage pipes hidden behind the walls and under the floor.

The Water Supply

The water supply system must assure you of pure water for drinking; it must supply a sufficient quantity of water at correct operating pressure at any outlet in the system; and it must furnish adequate hot or cold water, as required.

Water may be pumped from your own well directly into your home's plumbing system. Or it may come from a municipal or privately operated "waterworks" which purifies and distributes water through mains and branches. In the latter case, a connection is made at the street, where the main passes your house. Near this connection, a main shutoff valve is installed. This valve, which usually belongs to the water company, is opened and closed with a long-handled wrench called a "street key." Your water supply system starts at the valve.

If you are tied into a main system, the water will be piped through a meter, which may be located either outside or inside the house. Another valve is located where the piping enters the house, at the low part of the line. Past this valve—which, like the street valve, can be used to shut off all water in the house—the pipe becomes the cold water main line. Outdoor spigots, which use only cold water, can branch off this line. If a water softener is part of your home's system, the cold water main line should be tied into it. From the softener, the main line is split, with one pipe leading to the water heater. The cold water main then branches to the various fixtures and appliances: boiler, tub, lavatory, sink, toilet, washing machine. A parallel hot water main line also has branches to those fixtures requiring hot water (boiler and toilet need only cold water). Shutoff valves should be located in each branch line and in main lines wherever cut-off might be required.

Concealed in the wall at every branch line that terminates in a faucet is (or should be) an air chamber. This is a vertical, air-filled pipe that cushions onrushing water and absorbs pressure when the faucet is turned off. Without an air chamber, an effect called "water hammer" would cause pipes to chatter and even burst (see Chapter 9).

Finally, fixture supply lines lead the water from the branch lines to the individual fixtures. These should be fitted with shut-off valves (supply stops) to facilitate repairs when necessary.

Fixtures in the Plumbing System

Fixtures are located where you actually use the water. Even the faucet on the outside of your house, where you attach a hose for watering the lawn or washing the car, can be considered a fixture. Like all fixtures, it serves a specific purpose and must have certain features in order to perform its job; in this case, the faucet must be threaded so that you can attach the hose.

Similarly, shower fittings must be designed to mix hot and cold water to a desired temperature, and a toilet must be able to remove and replace a quantity of water in a relatively short amount of time. Always consider the purpose of a fixture or fitting when you buy a new one. And never skimp on quality. A few pen-

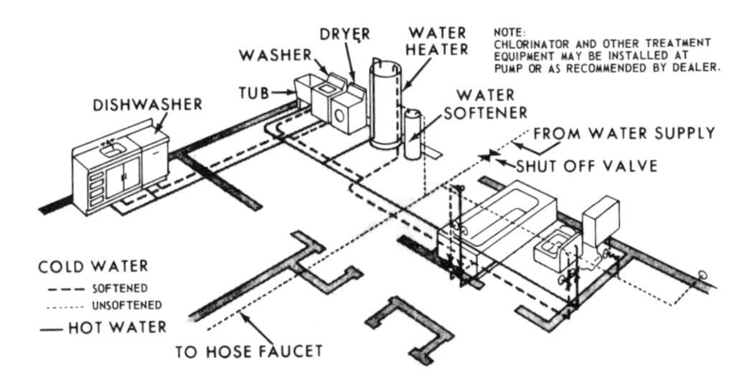

Above: *A typical water supply piping layout for a one-floor house.* Below: *The drainage system layout for the same house.*

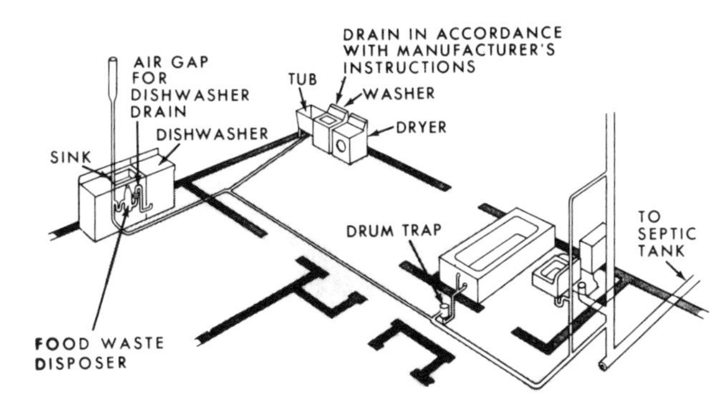

nies saved on fixtures can mean many dollars spent on repairs later—not to mention the aggravation when things go wrong.

Unlike the outside faucet, most indoor fixtures not only supply water but also provide for its removal after it has served its purpose. The water is usually removed through a receptacle connected to the drain-waste-vent part of your home's plumbing system.

Drainage

What goes up (or in) must come down (and out), and so it is with the water in your home's plumbing system. As the water supply starts at the well or the street and proceeds to the fixtures, the drainage starts at the fixtures and proceeds to the street and public sewers or to a private disposal system—by a different route, of course.

The drainage complex is more appropriately called drain-waste-vent (DWV), which best describes its multiple role. For health reasons and proper functioning, it must conform to certain simple but vital requirements:

- Pipes must be carefully and tightly fitted so that sewer gases cannot leak out.
- Vents must be provided to carry off the gases to where they can do no harm.
- Each fixture that has a drain should be provided with a suitable water trap to prevent gases from backing up into the room (the exception is a toilet, which has a built-in trap).
- "Re-vents" (bypass vents for air) must be installed wherever there is a possibility of siphoning water from a fixture trap. In addition, re-vents are sometimes required by local plumbing codes.

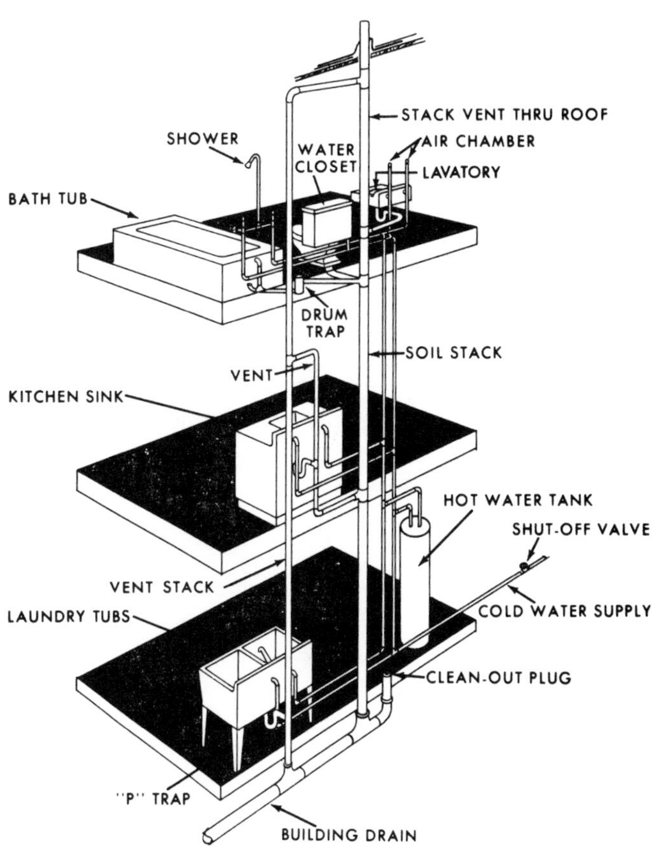

Typical plumbing arrangement for both the supply and drain-waste-vent system of a two-story house with a basement.

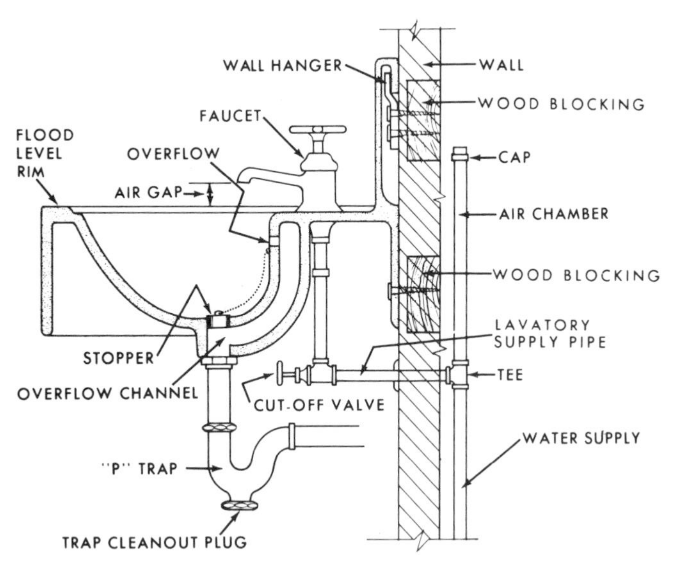

A bathroom sink, or lavatory, installation, showing supply and drainage piping. The air gap at the faucet prevents dangerous back-siphoning of waste water into the supply line. The air chamber in the wall prevents noisy "water hammer."

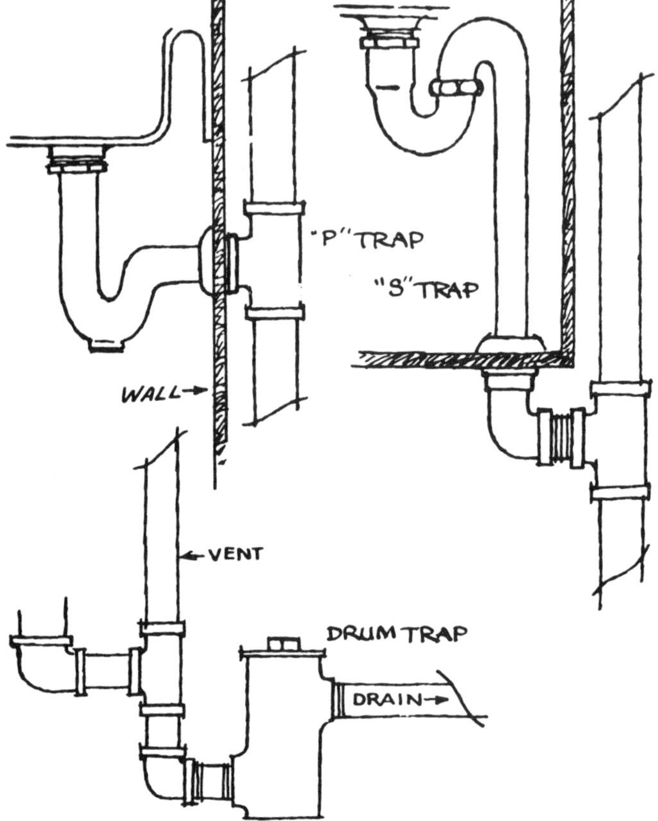

Different types of traps are used with fixtures. The drum trap is normally installed near a bathtub drain with its cover flush with the floor.

- Drainage pipes must be pitched or sloped for the downward gravitational flow of the water, all the way to final disposal.

The first step along the path that water takes to get out of the house is the fixture drain. This is where the trap is located (unless, as with toilets, it is built into the fixture). The trap is a water-filled P, J, S, or drum fitting (see illustrations) that guards against escape of gases into the house. From the fixture drain, the water goes into a branch drain or waste pipe, which slopes away from the fixture and leads the water to the soil (or main) stack. (There may also be smaller, secondary stacks in the drainage system; toilets must always drain into the main stack.)

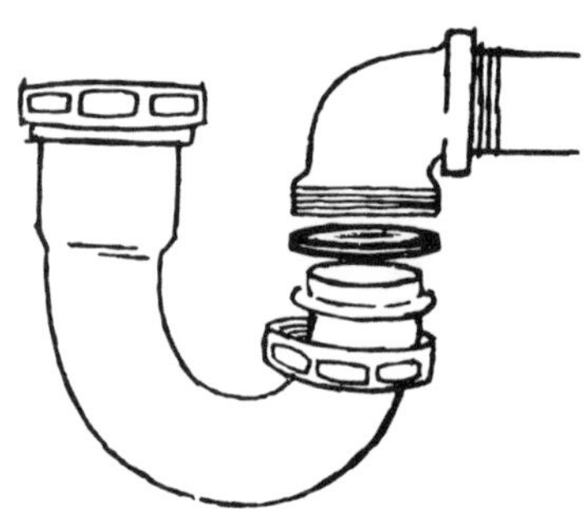

A P-trap assembly. Water in the trap prevents sewer gases in the drain from backing up and escaping into the house.

The upper portion of the soil stack is the vent, which goes up through the roof, allowing gases to escape to the outside and admitting air so that traps will not be siphoned dry by downrushing water. Re-vents are provided where a waste line cannot be vented directly into the stack or where the plumbing layout requires a bypass between a branch drain and the vent portion of the stack.

The soil stack carries the waste water to the building drain, which in turn carries it to the final place of disposal (a public sewer or your private septic tank and leaching field). At the foot of the soil stack, as well as at the foot of secondary stacks, there should be a cleanout. This allows you access to clear any obstructions that may clog the drain and waste pipes (diapers and sanitary napkins are not uncommon, although general household wastes can also accumulate and cause problems).

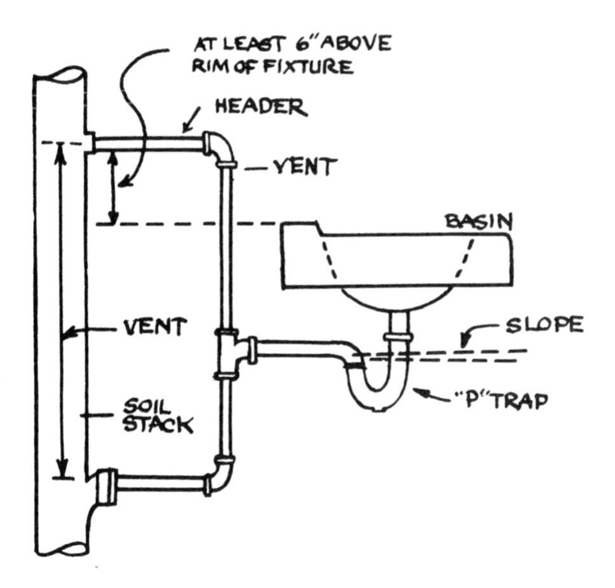

Fixtures are vented to carry off dangerous gases and to permit water and waste to travel freely through waste pipes without backing up into other fixtures.

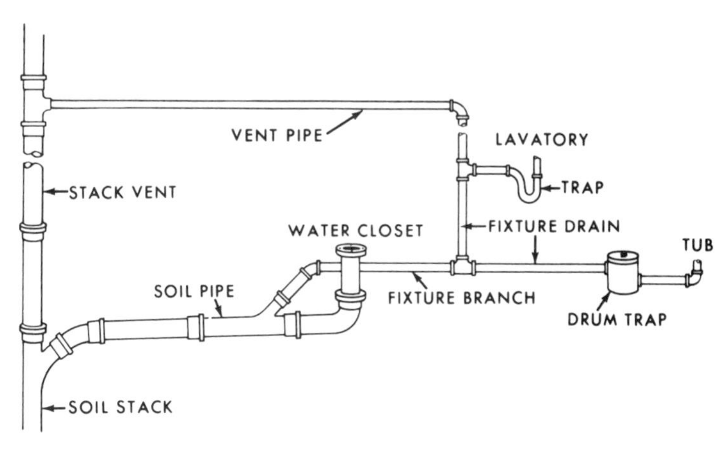

Venting a group of bathroom fixtures. The vent pipe is connected to the soil stack above the highest fixture drain (in this case, that of the lavatory).

Make a "Road Map" of Your Plumbing System

It is a good idea to sketch a simple "map" or diagram of your plumbing system. This will help you to pinpoint problems when they occur, and it is especially important if you finish off all or part of the basement for a recreation room or some other purpose—you may conceal the plumbing pipes in the process. A "map" will also help when you want to add new fixtures; for example, a backyard shower near your swimming pool.

Start where the water line enters your basement. Trace the line and its several branches until they disappear into the floor overhead. Then go upstairs and note where they reappear out of the wall and are connected to the various fixtures. Note the locations of all valves; then, when a problem arises and repairs must be made, you will be able to shut off only the necessary parts of the system. Also draw on your diagram the most probable routes of the piping through the walls from the basement to the fixtures.

Similarly, note the locations of the soil stack, secondary stacks, and building drain in the basement. Draw in the probable connecting lines that are concealed in the walls. Pinpoint the traps and cleanout plugs. Again, this will help when a drain becomes clogged and water is overflowing in your bathroom.

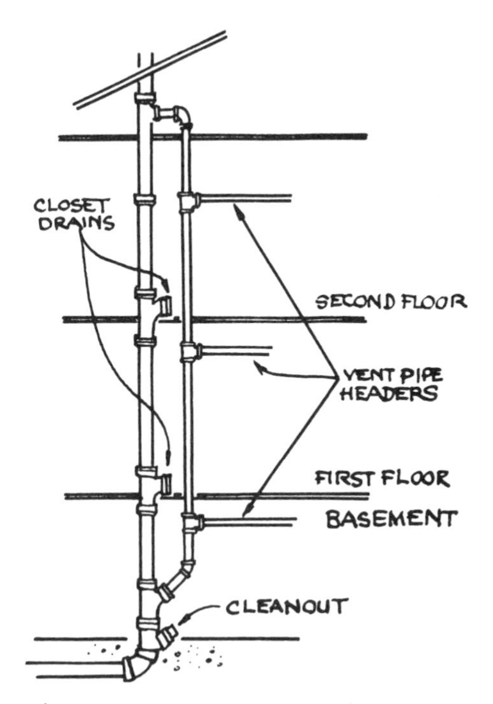

The main vent stack runs from the building drain in the basement floor up to the vent through the roof. Toilets must drain into the main stack; other fixtures may drain into secondary stacks.

The Plumbing Code

You don't need anybody's permission (just a little self-confidence) to change that faulty faucet washer or unclog that overflowing toilet. But when it comes to more ambitious projects —altering or adding to your home's plumbing system—there are most likely some stringent municipal regulations and restrictions governing what you can and can't do. Before you begin such a project, you should check with your local town hall.

Plumbing codes vary from one area to another. Some seem very archaic and restrictive, others follow the more up-to-date National Plumbing Code, and still others are willing to recognize and allow the use of the most modern materials and fittings. In some localities, a licensed plumber must be employed for all but the most simple projects; others allow the do-it-yourselfer fairly free rein. Almost all require inspections by municipal officials at various stages of the project. Make sure you get a copy of your local plumbing code early in the planning stages of your plumbing project; read it carefully and make sure you understand it—then conform to it rigidly. It is for your own protection.

Modern plumbing codes pay particular attention to the DWV system. In some older plumbing installations, improper layout could cause backup of waste water into the supply system, with obviously dangerous potential. If you live in a home with antique plumbing, you would be wise to get a copy of the plumbing code—even if you have no current plans to update the system—and to check out your piping for possible hazards.

2
Tools of the Plumbing Trade

One of the marks of a true professional—or a proficient amateur—is thorough familiarity with the tools of the trade. Without the proper tools, and without using them properly, a plumbing job simply cannot be done right.

For the most elementary plumbing repairs, such as changing a faucet washer, you need only the most elementary tools: a screwdriver and a wrench or pliers. But as you become more confident of your understanding of the plumbing system, you will probably want to undertake more major repairs and renovation tasks. For such projects you will have to expand your arsenal of tools. While there are many special, single-purpose plumbing tools, a selection of certain basic items will allow you to do most jobs.

Before you invest in pipe-working tools, note whether your system has copper tubing with soldered connections or whether it has steel pipes. Each type calls for a separate set of tools. Working with drainage pipe requires still other tools.

The Emergency Toolbox

Forearmed is fortunate, so be prepared for plumbing crises before they occur. If your toilet overflows while you're watching the late-late show, you won't be able to run to the hardware store for supplies.

You probably already have many of the small tools you will need: screwdrivers (including Phillips-heads, if your faucet handles are held on by this type of screw), an adjustable wrench or locking pliers, and electrician's tape. These, plus a supply of faucet washers in assorted sizes, will cure most dripping-faucet ailments. You might also want to invest a dollar or so in a seat-dressing tool, in case a new washer doesn't do the job.

Clogged drains often can be cleared with drain cleaner of either the liquid or dry type. A drain cleaner should always be kept on hand near the kitchen sink. And no house should be without the familiar "plumber's friend," or force cup—a rubber suction cup on the end of a handle. The ball-type plunger, with a rounded bottom rather than a large flat opening, exerts more force and is especially effective for unclogging toilets. An auger or "snake"—spring steel or coiled wire with a small metal handle—may be needed to remove particularly stubborn obstructions in waste and drainage pipes.

With these tools at hand, you should be able to cope with most common plumbing emergencies. Except for emergency tools, only buy tools as the need for them arises. The do-it-yourselfer often buys on impulse and builds up a collection of gadgets that may be lovely to look at but which will serve only to adorn the PegBoard wall behind the workbench.

Drain cleaner should always be kept in your emergency toolbox. Used according to manufacturer's directions, it will cure many clogging problems.

The old standby, the "plumber's friend," is another problem solver. It's a good idea to station one unobtrusively in every bathroom.

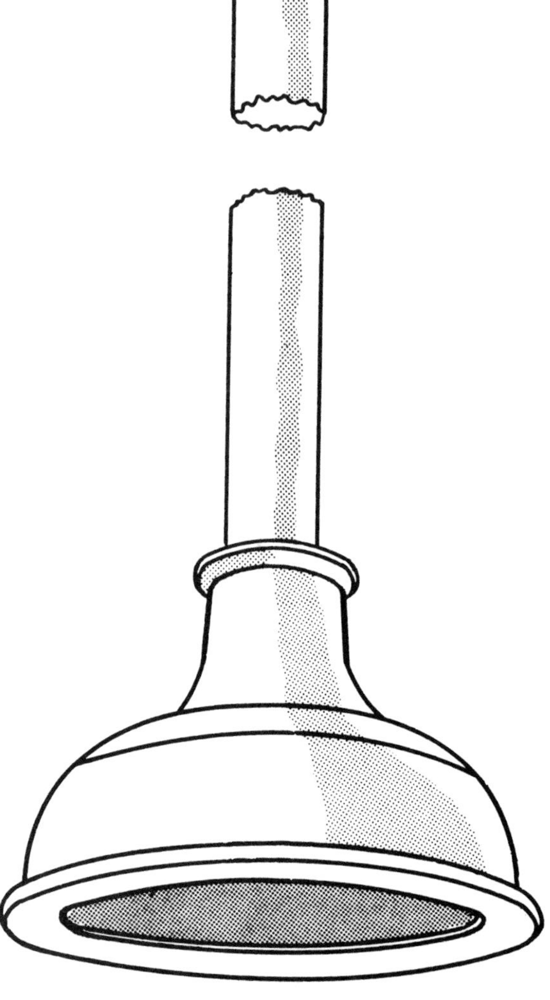

The type of force cup illustrated here has a large, flat opening at the bottom of its rubber plunger. Another type, with a rounded bottom, is even more effective, especially when clearing a clogged toilet.

Fixed Wrenches

As you tackle more ambitious plumbing projects, your tool collection will grow. Basic to the plumber's craft, wrenches are divided into two broad classifications: fixed and adjustable. Fixed wrenches (open-end and box) are of a specific size and cannot be adjusted. Their plumbing uses are somewhat limited, although a good set of graduated sizes is always handy to have in your toolbox. For working in tight places, closet spud wrenches—special thin open-end types—are particularly helpful.

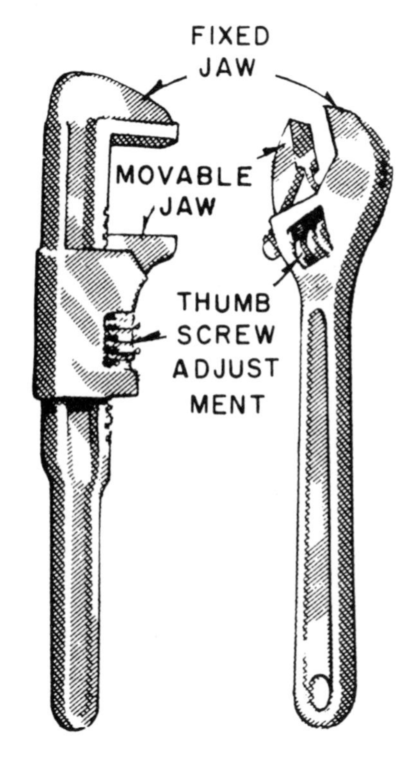

Both the monkey wrench (left) and adjustable open-end wrench (right) have a fixed jaw and a movable jaw. The movable jaw is controlled by a thumbscrew.

A special purpose tool is the packing-nut socket wrench, which comes in various sizes to fit tub and shower valves. Hexagonal in shape and with a hollow core, the wrench fits over the faucet stem to remove the valve packing nuts and stem assembly. Since these nuts and stem assemblies are made of brass, ordinary wrenches should not be used on them. Applying too much pressure will bend or break the fittings, which are practically impossible to replace. About the only thing you can do in such a case is tear out the wall and install new valves. It is far cheaper to purchase the right-size socket wrench.

Adjustable Wrenches

Adjustable wrenches can be opened or closed. Within the limitations of the individual wrench, they fit nuts, bolts, and pipes of various sizes. There are many different types of adjustable wrenches, several of which are used in plumbing.

Adjustable Open-End Wrench: This type of wrench is used on square or hexagonal nuts when working with the interior parts of faucets and valves. For most minor plumbing jobs, a 12-inch wrench is probably the best choice (the size refers to the length of the tool; however, the jaws of larger wrenches do open more widely). It is also handy to have an 8-inch or 6-inch wrench for smaller work.

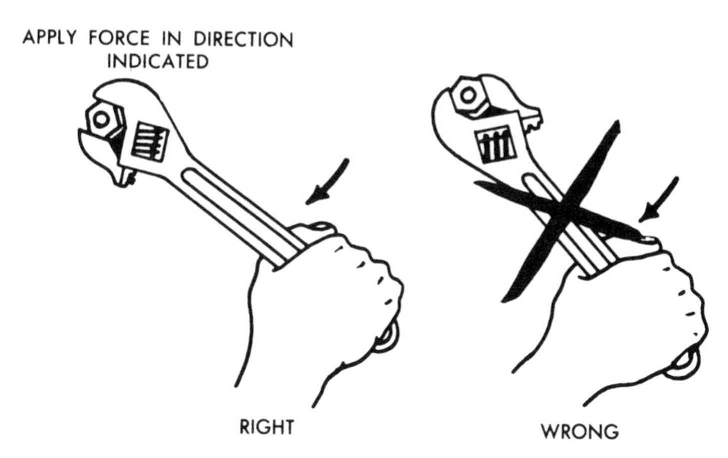

When using an adjustable wrench, apply force in the direction of the movable jaw.

Always place the wrench on the nut so that you pull the handle toward the side with the adjustable jaw. This will prevent the jaw from opening and slipping off the nut. Be sure that the nut is all the way into the throat of the jaws and that the jaws are adjusted to fit snugly around the nut. Otherwise, the wrench may slip off, damaging the nut or valve—and bruising your knuckles.

Monkey Wrench: This type of wrench is used for the same purposes and in the same way as the adjustable open-end wrench. Turning force should be applied to the back of the handle (the side of the wrench opposite the jaw opening).

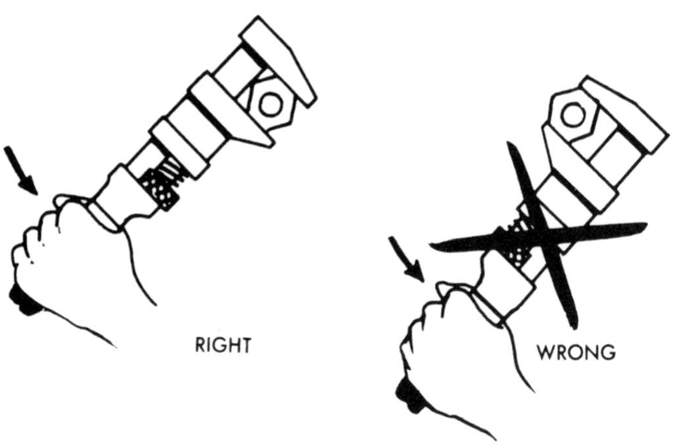

When using a monkey wrench, apply force in the direction of the movable jaw.

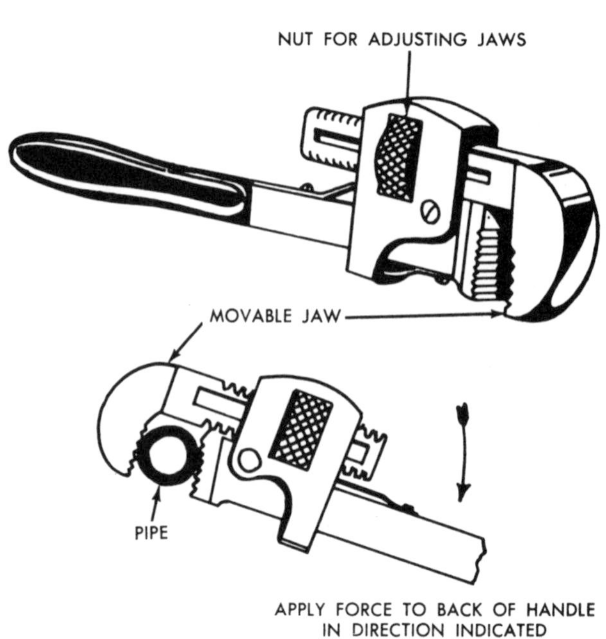

On both pipe and Stillson wrenches, force should be applied to the back of the handle.

Pipe Wrench, Stillson Wrench: Used for working with iron pipe, these wrenches have hardened steel jaws that provide excellent bite and grip. The main difference between a pipe wrench and a Stillson wrench is that the Stillson wrench has a separate housing in which the adjusting nut operates. In both types, the movable jaw is pivoted, permitting a gripping action. Since there is little difference between the two wrenches, which one is used is a matter of individual preference.

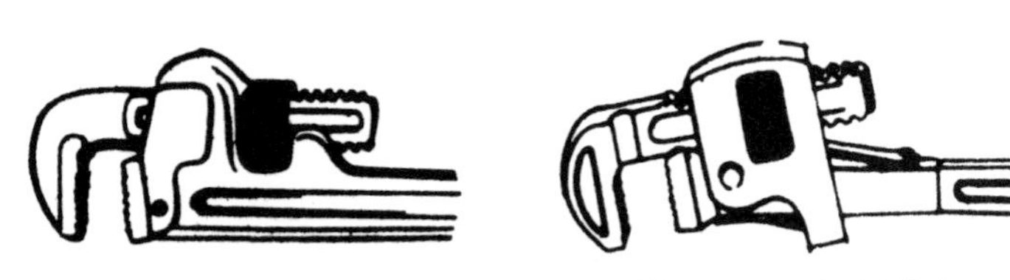

A pipe wrench (left) and a Stillson wrench (right) do the same job, but the adjusting nut of the Stillson wrench is in a separate housing.

Pipe wrenches come in many sizes. For small pipes, a 6-inch or 8-inch wrench is best. For pipes in the ½-inch to 1½-inch range, a 12-inch or 14-inch wrench should be used. For pipes between 1½ and 2 inches, the wrench should be 18 inches long.

Generally, two wrenches are needed: one for holding the pipe and one for turning the fitting. Since pipe wrenches work in one direction only, they should be placed on the work in opposing directions. Adjust the jaws so that the bite on the work is taken at about the

Generally, pipe and Stillson wrenches are used in pairs; one acts as a vise while the other turns the pipe or fitting.

center of the jaws. Always turn the wrench in the direction of the opening of the jaws, applying force to the back of the handle. The grip on the work is increased by adding pressure to the wrench handle.

The serrated jaws of pipe and Stillson wrenches bite into the metal being worked and always leave marks. For this reason, they should not be used for tightening or loosening nuts; nor should they be used on tubing or chrome-finished pipe that is easily marred.

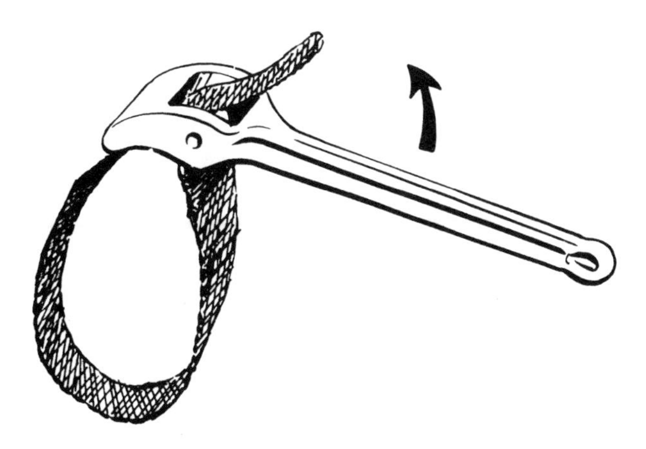

A strap wrench is used on pipe that might be damaged by the serrated jaws of a conventional wrench.

Strap Wrench: This type of wrench is recommended for working with brass, aluminum, lead, soft metal, or plastic pipe. It consists of a handle and a heavy, webbed strap that is looped around the pipe in the opposite direction to that in which the pipe is to be rotated. The strap is then passed through a slot in the handle and drawn up tightly. As the handle is pulled, the strap tightens further, gripping the pipe and turning it.

Chain Wrench: For pipes with large diameters such as cast-iron drainpipes, a chain wrench should be used. This consists of a forged steel handle, to which a length of heavy sprocket chain is attached. When looped around the pipe, the chain acts as a jaw, gripping the entire outer circumference of the pipe. The tool works in one direction only, but the handle can be backed partly around the work and a fresh hold taken without freeing the chain.

The chain wrench is used for larger steel pipe or cast-iron pipe.

Basin Wrench: This is a specialized wrench. It is used to remove or tighten nuts on faucets and spray attachments in hard-to-reach spots under sinks.

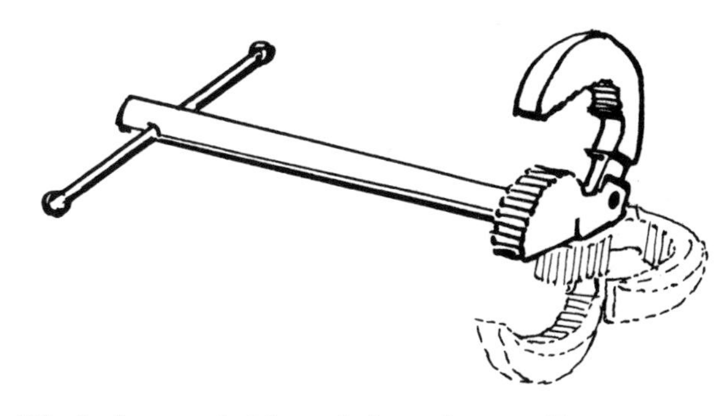

The basin wrench (shown in its various working positions) is used on hard-to-reach nuts under sinks.

Pliers

Pliers often are useful for holding and gripping, but they should never be substituted for wrenches. Do not use pliers to turn nuts—it takes only a few turns to damage the nut. In addition to the common household pliers, there are some specialized types that are handy for plumbing operations.

Water-Pump Pliers: This tool was originally designed for tightening or removing water-pump packing nuts. One of its two jaws is adjustable to seven different positions. The inner surface of both jaws consists of a series of

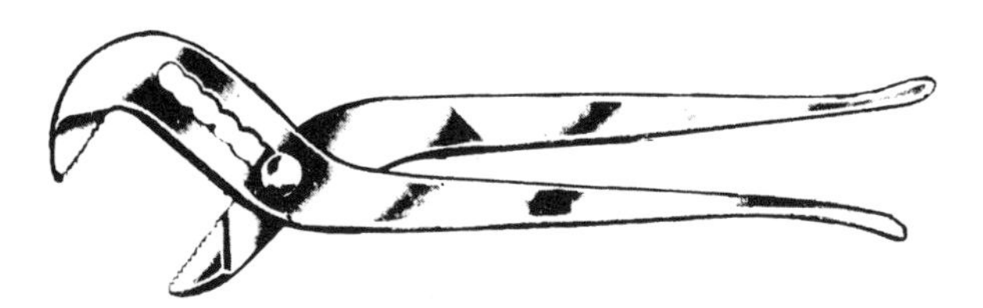

Water-pump pliers have a series of grooves for adjusting to cylindrical objects.

coarse teeth formed by deep grooves. Such a surface is especially good for grasping cylindrical objects.

Channel-Lock Pliers: These are similar to water-pump pliers, but the jaw-opening adjustment is different. Lands on one jaw fit into grooves on the other. Channel-lock pliers are less likely to slip from the adjustment setting when gripping an object, but they should be used only where it is impossible to use a wrench or holding device because they could do damage to a surface.

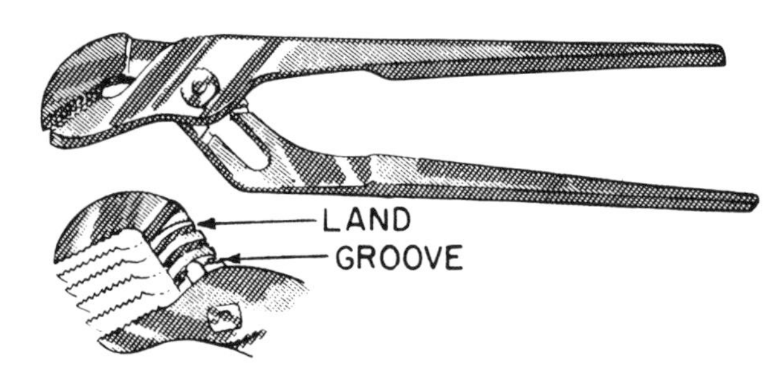

Channel-lock pliers are designed to lock when gripping an object.

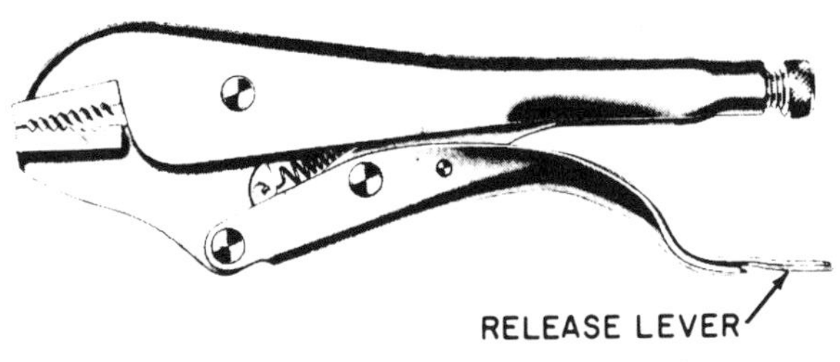

Vise-grip pliers are adjusted by a knurled screw on the handle. Many types also include a release lever.

Vise-Grip Pliers: These pliers are very useful when working with small-diameter pipes. They have an advantage in that you can clamp them onto an object and they will stay, leaving your hands free for other work. The jaw opening is adjusted by turning a knurled screw at the end of the handle; then the tool is clamped onto the pipe or other object. It is locked in position by pulling the lever toward the handle.

Vise-grip pliers should not be used on nuts, bolts, or tube fittings because the teeth in the jaws tend to damage the surface to which they are clamped.

Plumbers' Vises

Plumbers' vises differ from ordinary bench vises or machinists' vises in that they have serrated jaws. These jaws grip the pipe and prevent it from turning. In selecting a plumbers' vise, it is important to consider how securely the tool holds the pipe and whether or not it is large enough to handle all the sizes of pipe with which you will be working.

Yoke Vise: This vise is specifically designed for holding pipe or round stock, and it is preferred by most plumbers. Usually bolted to a bench, it has V-shaped jaws that grip the pipe from both the top and the bottom. The upper jaw is on a hinged yoke, which can be raised to allow you to position the work. The jaw is brought down and locked by turning an adjusting screw.

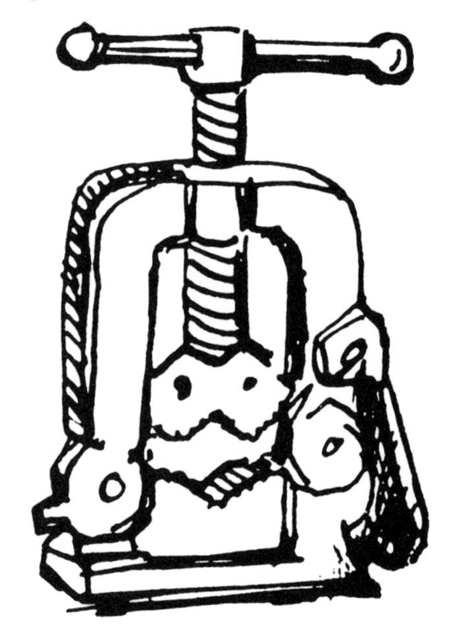

A yoke vise is usually mounted on a bench or stand.

Combination Vise: Basically a machinists' vise, this tool has integral jaws for holding the pipe. It can be used in emergencies, but it is not recommended for heavy-duty work.

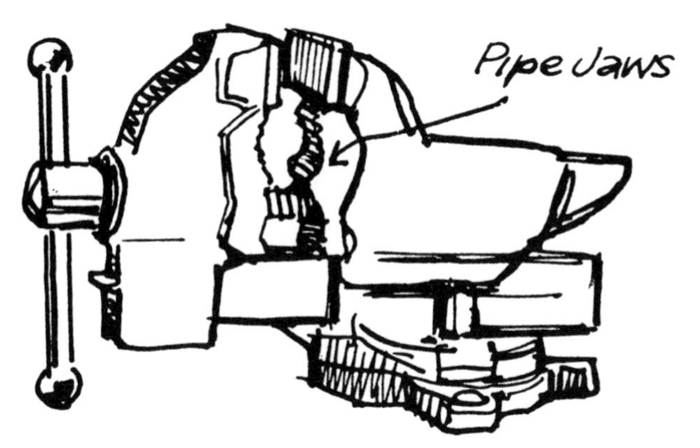

A combination vise can be used in emergencies, but it is not recommended for general pipe work.

Chain Vise: This type of vise has a fixed, V-shaped, lower jaw with teeth. The pipe is laid on this jaw, and a bicycle-type chain is fastened to one side. After the pipe is in position, the chain is brought over it and locked securely in a slot on the other side.

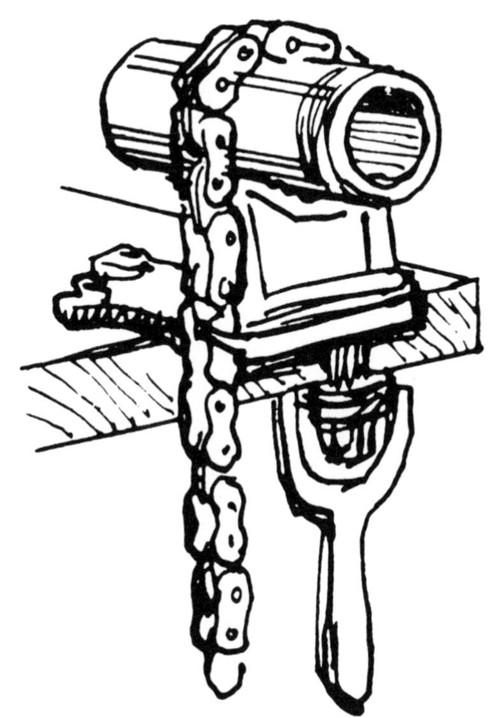

A chain vise can be used for steel or cast-iron pipe.

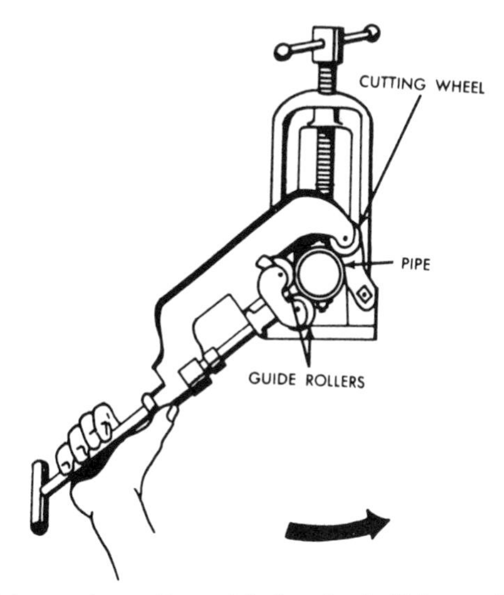

Using a pipe cutter, with the pipe held in a yoke vise.

Pipe and Tubing Cutters

A hacksaw can be used to cut pipe and tubing, but it is almost impossible to make a straight, clean cut this way—and it is hard work, especially when you are sawing through steel pipe. Using a pipe cutter will give you a more accurate and quicker job.

Most pipe cutters have a single cutting wheel and two rollers. To use this tool, start the cutter over the pipe and keep turning it, tightening the handle slightly as you go. Lubricate both the cutting wheel and the pipe with thread-cutting oil.

A tubing cutter; pointed device is for reaming the cut ends of copper pipe.

A tube cutter should be used to cut copper tubing. This is similar to a pipe cutter, but it is smaller and lighter. A vise is not ordinarily used to cut copper tubing, since both the tubing and the cutter can be hand-held. No cutting oils are required, but do not feed the cut too quickly or you may flatten the tubing.

Reamers and Files

Whenever pipe or tubing is cut, both the inside and the outside edges retain burrs. Those on the outside can interfere with the threading of the pipe or the joining of tubing; they also pose a threat to your hands while you are working with the pipe. Burrs on the inside edge cause obstructions within the pipe.

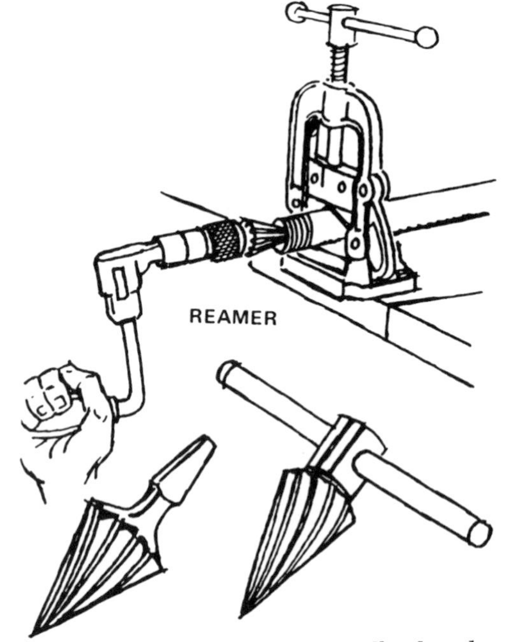

Spiral-type reamers, one with a T-handle, the other for use in a brace.

Use a flat file to remove burrs from the outside edge of the pipe. A spiral-type reamer is best for removing the burrs inside the pipe or tubing.

Pipe Threaders

Generally, you can buy pipe cut and threaded to order. If, for some reason, you decide to thread your own, you will need dies in the size of the fittings the threaded pipe will connect to. These dies are held in a die stock that has long handles to turn the tool on the pipe.

When cutting threads, it is necessary to hold the pipe securely in a vise. The thread should be lubricated frequently with cutting oil. The length of the thread depends on the size of the pipe.

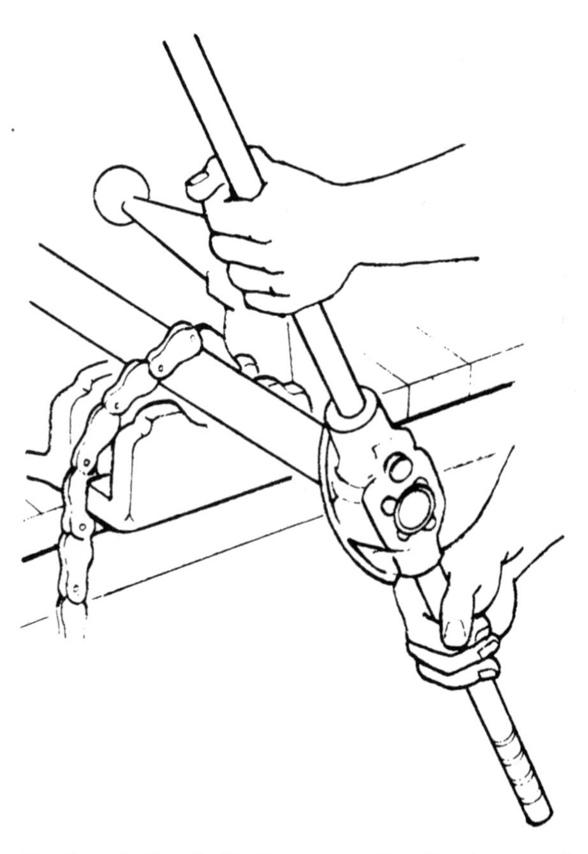

To thread pipe, hold the proper-size dies in a stock with handles for turning.

Diameter of Pipe (inches)	Length of Thread (inches)
1/4	5/8
3/8	5/8
1/2	13/16
3/4	13/16
1	1
1 1/4	1
1 1/2	1
2	1 1/8
2 1/2	1 9/16
3	1 5/8

Flaring Tools

Many valves and other fittings to which tubing is attached are of the "flare" type, which requires that the tubing be flared or turned outward at the end.

The most common type of flaring tool con-

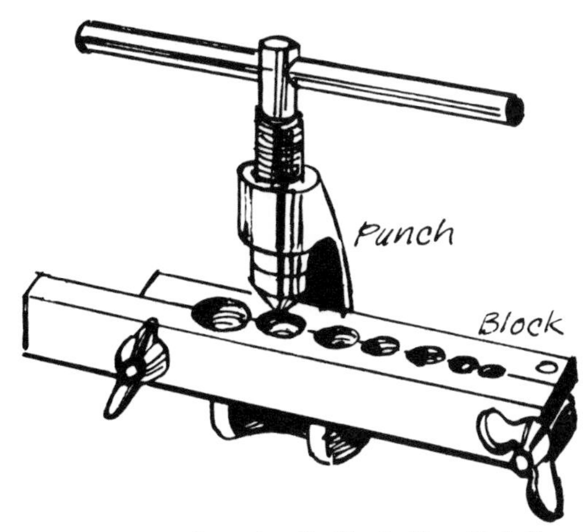

Flaring tool consists of a die block (in which the tubing is clamped) and a yoke (which includes a handle with a punch on the bottom that forms the bell-shaped flare).

sists of a split die block with holes for various sizes of tubing, a clamp to lock the tubing in the die block, a yoke with a compressor screw that fits over the die block, and a cone at the base of the screw that forms a 45° flare or bell shape on the end of the tubing. First, the sleeve nut of the fitting is slipped onto the tubing; then the tubing is clamped in the die block and the compressor screw is tightened to form the flare.

Propane Torches

For making solder connections in copper plumbing lines (see Chapter 4), a propane torch is the best tool to use. It is easier to handle than the larger blowtorch—and much safer too. When the propane is exhausted, the soldering head is simply transferred to a fresh tank.

Tools for Cast-Iron Pipe

To cut cast-iron drainage pipe, you will need a hacksaw, a cold chisel, and a hammer. Select a 1-pound ball peen hammer or a short-handled sledge hammer for this job. Never use a claw hammer, which is designed for carpentry only.

You also will need a yarning iron to drive oakum into the joints, a blowtorch and melting pot to melt lead, and a cast-iron ladle for pouring the hot lead into the joints. Inside and outside calking irons are used to finish the joints.

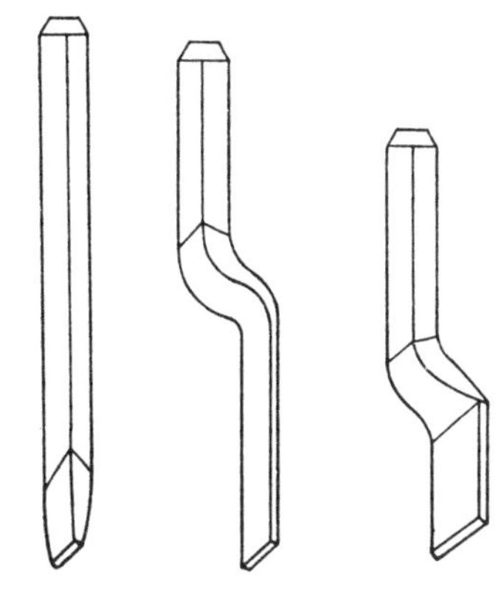

Among the tools needed for working with cast-iron drainpipe: cold chisel (left), yarning iron (middle), and calking iron (right); the latter is made in both inside and outside configurations (see Chapter 6).

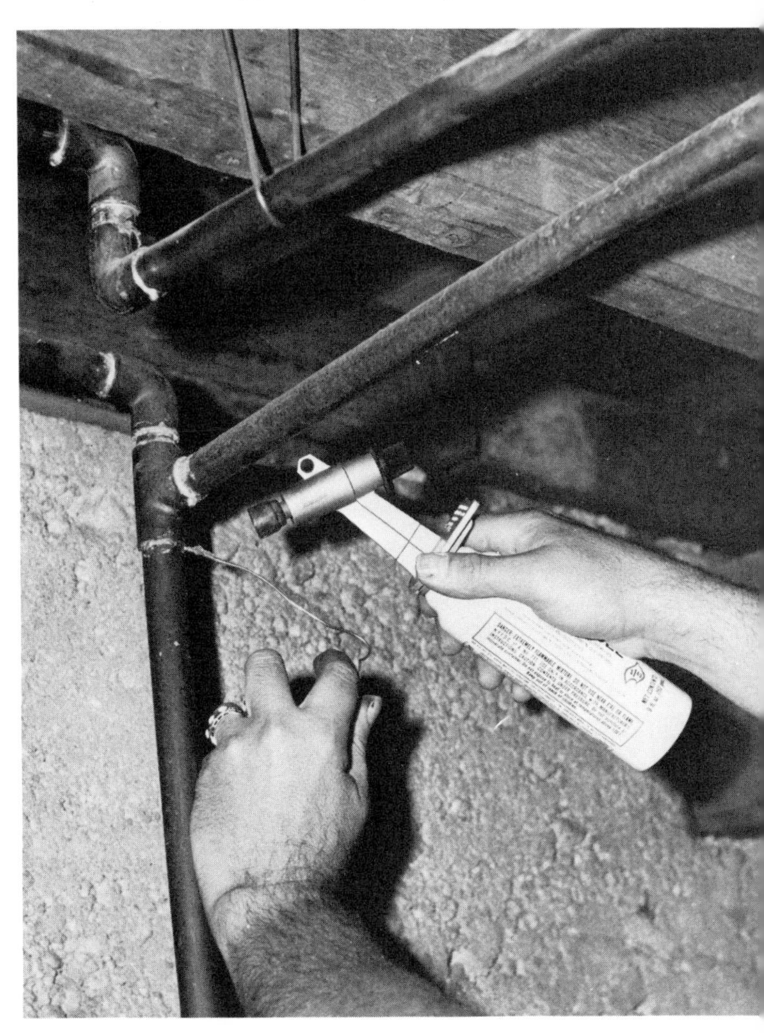

A propane torch is the best and safest tool to use when soldering copper pipe joints.

3
Water Supply Pipes and Fittings

"Getting there is half the job" of your plumbing system. The pipes that carry the water from where it enters your house to the faucet (or to wherever else it will be used) may be made of galvanized steel, copper (either rigid, soft, or flexible) or plastic.

Galvanized Steel Pipe

Galvanized steel pipe is sold in sizes ranging from ⅛ inch to 6 inches. Most residential water supplies use ½-, ¾-, or 1-inch pipe.

Steel pipe is tough, durable, and more economical than copper. But it is generally more difficult to install because of its rigidity and because of the threading it requires. In a tight situation, where it is difficult or impossible to swing a wrench, copper or plastic pipe is probably a better choice.

Sizes for Steel Pipe

Nominal size in inches	Outside diameter in inches	Inside diameter in inches	No. of threads per inch
¼	.540	.364	18
⅜	.675	.493	18
½	.840	.622	14
¾	.050	.824	14
1	1.315	1.049	11½
1¼	1.660	1.380	11½
1½	1.900	1.610	11½
2	2.375	2.067	11½
2½	2.875	2.469	8

Copper Pipe

Copper pipe is available in three types: hard, soft, and flexible tubing. All types come in ⅜-, ½-, and ¾-inch sizes (for in-house use) and the hard comes in ¾-inch and larger sizes (for underground use). Since water flows through copper tubing with less resistance than through steel pipe, a smaller diameter line often can be used when making a replacement. Special adapters make it possible to combine copper tubing with a system of threaded steel pipes.

Hard, or rigid, copper tubing is assembled with solder-type fittings. It makes a very permanent trouble-free installation. Where long lines are exposed, such as in the basement, rigid copper tubing gives the best appearance.

Soft copper tubing is easier to install than the hard, and flexible tubing is even easier; both types can turn corners without elbows, and it is a simple matter to make slight bends to adjust for any inaccuracies. Either solder-type fittings or flare-type fittings may be used with soft copper tubing. Flare-type fittings are attached simply by being tightened and are very easy to assemble. However, they are not as durable as soldered joints and should never be used inside walls or in other places where they cannot be reached easily for repair or replacement. Solder, flare-type, or compression fittings can be used with flexible tubing.

Copper pipe is manufactured in three grades. Type K, which may be hard, soft, or flexible, is the heaviest because it has the thickest walls. It is used primarily in commercial work. Type L, which may be hard, soft, or flexible, is lighter than type K and is used for the most part in residential water lines. Type M copper piping, only available as hard tubing, is used for light residential lines. Since some municipal codes do not permit the use of type M copper pipe, check your local code before using it. Soft copper tubing (types K and L) is sold in 30-foot and 60-foot coils. Hard copper tubing (types K, L, and M) is sold in 12-foot and 20-foot lengths. Flexible tubing is sold only in 60-foot coils.

Galvanized steel pipe comes in 21-foot lengths; you can have your dealer cut it to shorter lengths as needed. Rigid copper tubing comes in various lengths, while flexible copper tubing is purchased in rolls of 60 feet.

Sizes of Types K, L, and M Copper Tubing

Nominal size in inches	Outside diameter	Inside diameter		
	Types K, L, M	*Type K*	*Type L*	*Type M*
⅜	.500	.402	.430	.450
½	.625	.527	.545	.569
¾	.875	.745	.785	.811
1	1.125	.995	1.025	1.055
1¼	1.375	1.245	1.265	1.291
1½	1.625	1.481	1.505	1.527

Plastic Pipe

Technological advances in the plastics industry have made many municipal building codes obsolete, so make certain that you check your local plumbing code before installing any plastic pipe.

Plastic pipe is easier to install than any type of metal pipe. It will not rust, rot, or corrode, and it is extremely light to handle. It can be cut with a hand saw—some types can even be cut with a sharp knife. Fittings are joined with a solvent cement, or simplified snap-on fittings can be used.

There are many different types of plastic pipe, but only three are commonly used by do-it-yourselfers. All three are suitable for underground as well as in-house installation, but only one can be used for hot water lines.

Chlorinated polyvinyl chloride (CPVC) plastic pipe can be used in both hot and cold water supply lines. It comes in semirigid lengths from 10 feet to 40 feet. A near-relative, *polyvinyl chloride (PVC)* pipe, comes in the same lengths. However, it can only be used for cold water, since heat will soften the plastic. *Polyethylene (PE)* pipe is flexible and comes in long coils, making it especially useful for bringing water from a water main or well into the house.

Whatever type of plastic pipe you select, if the water it will carry is to be used for drinking, make sure that it bears the seal of the National Sanitation Foundation.

Many types of fittings are available for use with galvanized steel pipe.

Pipe Fittings

There's many a twist and turn 'twixt the water main and the faucet–and there are fittings to match every one of them. However, all pipe fittings have certain similarities. Fittings are ordered according to the size of pipe on which they will be used (a ½-inch elbow will take ½-inch pipe on both ends). Some of the most commonly used fittings are discussed below.

Nipples are used to add short sections to galvanized pipe lines. They come in all pipe sizes and in lengths from close (nipples that are threaded on both ends to a point where threads almost meet in the center) to as large as 24 inches.

Couplings connect pipes of the same size that normally will not have to be disconnected.

Reducing couplings connect pipes of different sizes that normally will not have to be disconnected. Some reduce pipe only one size (for example, from ¾-inch to ½-inch); others may reduce pipe several sizes (for example, from 1¼-inch to ½-inch).

Unions connect pipes that may have to be disconnected at some future time (for example, pipes leading into and out of the water heater). The fitting is in three parts. One part is attached to each of the pipes to be joined, then the third part (a special nut) pulls them together.

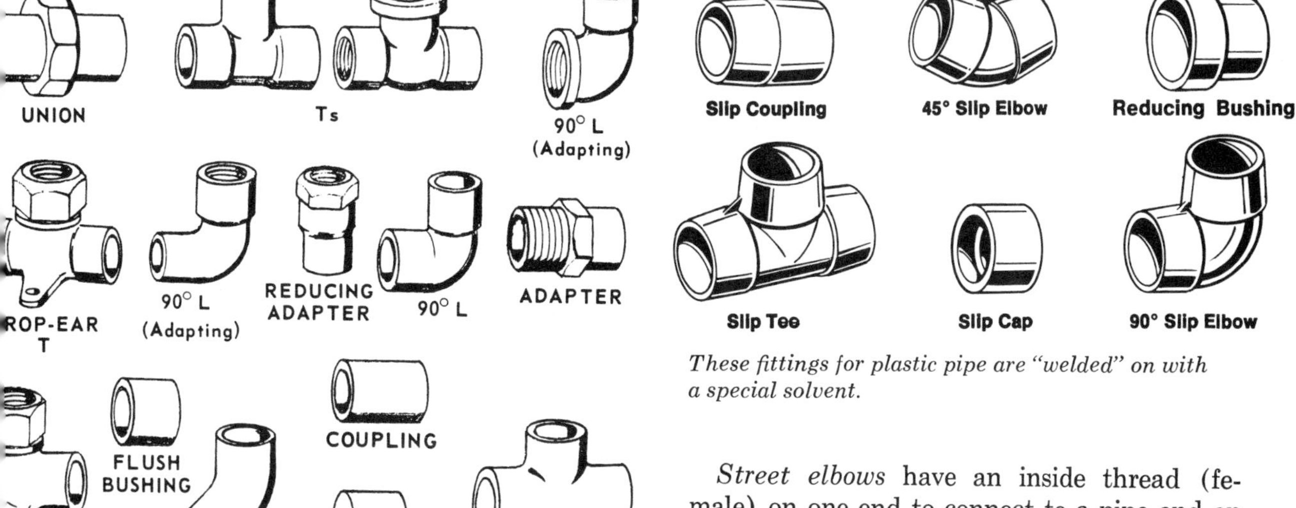

Soldered fittings, such as those shown here, are used on rigid copper tubing.

These fittings for plastic pipe are "welded" on with a special solvent.

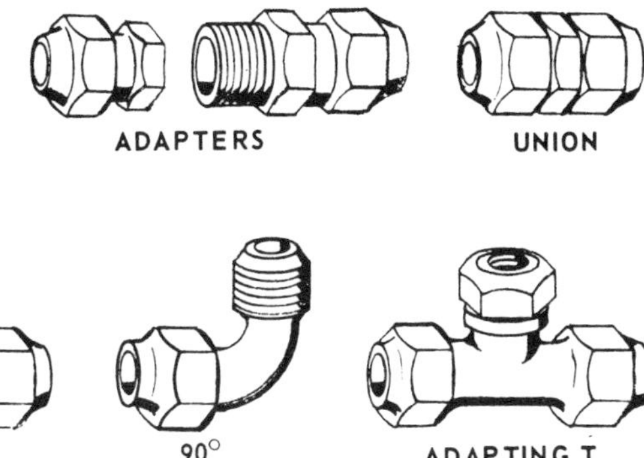

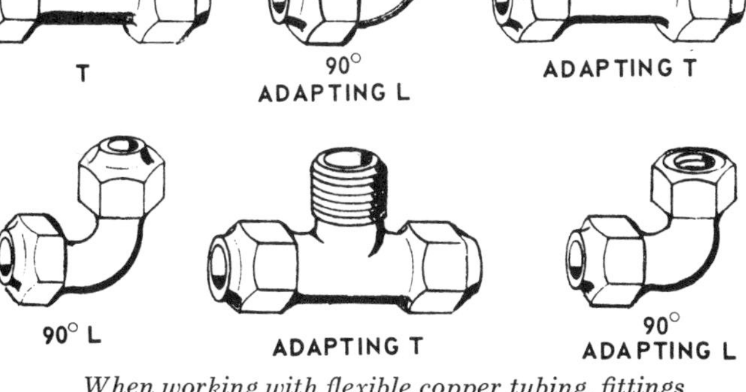

When working with flexible copper tubing, fittings may be soldered as with rigid tubing, or flare-type.

Bushings are inserted inside a coupling or other fitting to reduce the size.

Elbows change the direction of pipe at an angle of 90° or 45°.

Reducing elbows join two pipes of dissimilar sizes at an angle of 90° or 45°.

Street elbows have an inside thread (female) on one end to connect to a pipe and an outside thread (male) on the other end to connect to another fitting.

Tees join two pipes to a third at a 90° angle.

Reducing tees join three pipes of two or three different sizes. There are various combinations of such tees; for example, ¾″–¾″–½″, ¾″–½″–¾″, ¾″–½″–½″, and 1″–¾″–½″.

Crosses join four pipes at 90° angles to each other.

Plugs seal off the female end of fittings.

Caps close off the end of a pipe.

Valves

Valves are placed in water supply lines wherever it may be desirable to shut off the flow of water at times. Valves used in residen-

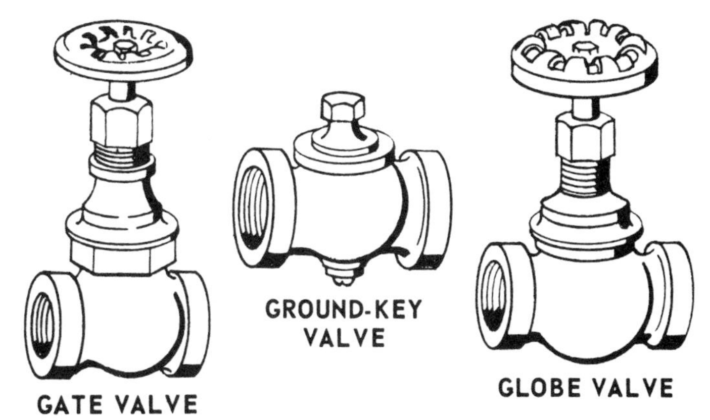

These types of valves are found commonly in the home's water supply system.

tial plumbing are usually of cast bronze. There are different types for different purposes. Since some are designed to restrict water flow even when open, it is important to select the right valve for the job you want it to do.

Gate Valve: This valve is used to completely shut off or open a water line. It cannot be used to control the volume of flow. It has a sliding wedge that is moved across the waterway, usually by a threaded spindle or stem.

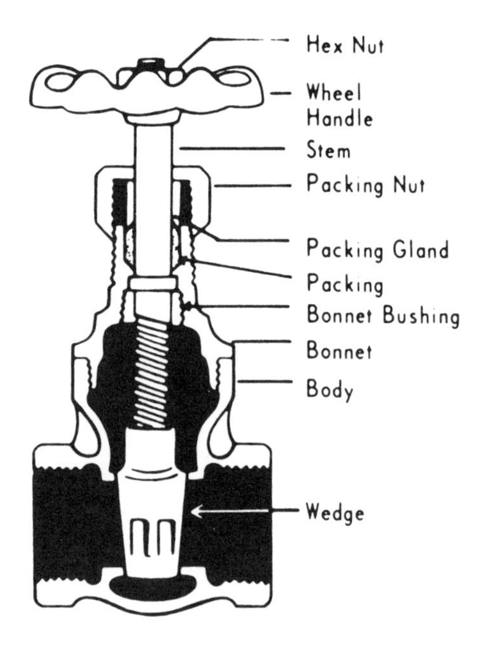

Inside a gate valve. When the wedge is fully opened, water flows almost unrestricted.

Since the flow of water is virtually unrestricted, full water pressure is allowed when the valve is open. However, dirt or sand in the seat may prevent the valve from closing fully. If this happens, or if the valve seat or wedge wears out, the valve should be replaced.

In residential water supply lines, gate valves are normally used only where the house line connects to the main.

Globe Valve: This type of valve is used where it must be opened and closed frequently and where water pressure is relatively high.

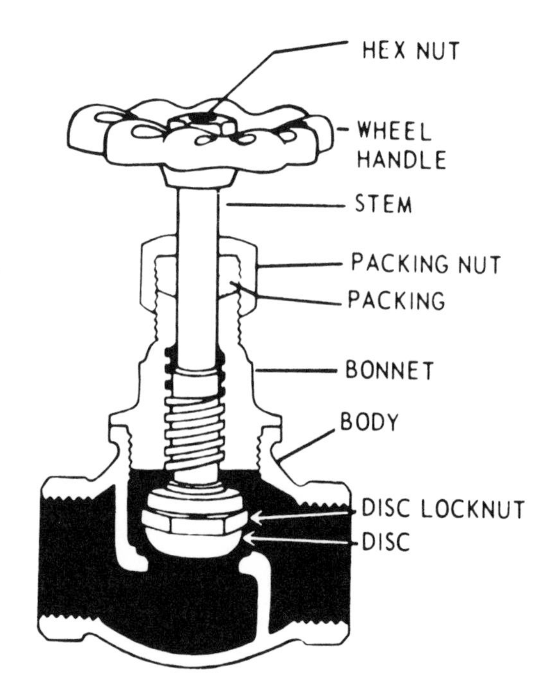

Construction of a globe valve. Note that partitions within the valve force water flowing through to change course several times, restricting the flow.

Because a globe valve constricts the waterway, it is often used to control the volume of flow.

The globe valve contains two chambers that are separated by a partition. Water flowing through the valve must change course several times from port to port (the intake and outflow openings), causing resistance. Turning the handle of the valve increases or decreases the opening of the passage.

The globe valve is the type most commonly found in residential water supply lines. Its design is simple, and it is fairly easy for the do-it-yourself plumber to service when something goes wrong.

Angle Valve: This valve is similar to a globe valve, but the ports are at right angles to each other. Since the 90° change of direction already offers resistance to water flow, the internal design of the angle valve does not include the globe valve's diverting partition. Since the water must change direction only once, its flow is freer than through a globe valve. When installed at a turn in the piping,

Angle valves are similar to globe valves; they are often preferred where a line must make a 90° turn.

an angle valve can be used instead of a globe valve and an elbow.

Key Valve: While commonly found in gas lines, key valves are sometimes used in water lines as well (for example, on supply lines for automatic washers). Inside, a tapered ground plug seats into a matched ground body. Outside, a lever usually controls the on–off action. A quarter turn opens or closes the line. Key valves, like globe valves, restrict the water flow.

Drainable Valve: A drainable valve (sometimes called a bleeder valve) should be installed to control the flow of water to any outside connection, such as the faucet to which you attach your garden hose, unless it is a special outdoor fitting—see page 28. Manufactured in both gate and globe types, drainable valves have a cap or screw on the side.

A drainable valve has a plug on the side that can be removed when it is necessary to drain the water line.

When the water to the outside is turned off for the winter, this cap or screw can be opened to drain off the water remaining in the line. This prevents the freezing and possibly the bursting of the pipe.

When installing this type of valve, make sure that the drain opening is on the nonpressure (outer) side of the valve seat.

Check Valve: A check valve allows water to flow in one direction only, operating automatically to prevent backflow. The most common type has a swinging gate which opens only one way to allow water through; pressure from the opposite side forces it shut.

Some communities require the installation of a check valve in a cold water line between the meter and the water heater. Check valves also are used to prevent water pumped to an overhead tank from flowing back when the pump stops.

Different types of check valves are manufactured for horizontal and vertical installation. Make sure you select the right one.

Vacuum-Breaker Valve: Similar in function to a check valve, a vacuum-breaker valve is used with washing machines, dishwashers, and other installations to prevent the siphoning of polluted water back into the supply system. When the pressure in the water supply line drops, air enters the valve to fill the space and to prevent backup.

Faucets

Faucets are simply valves placed at the end of water supply lines. There are two basic types for residential use: compression and noncompression. Each type has several variations for specific purposes. In older homes, Fuller ball and ground-key faucets also may be found.

Compression Faucet: This faucet regulates water flow by means of a lever, T-handle, or ball-handle attached to a threaded spindle. When the spindle is turned down, a washer or disc (attached to its lower end) is pressed

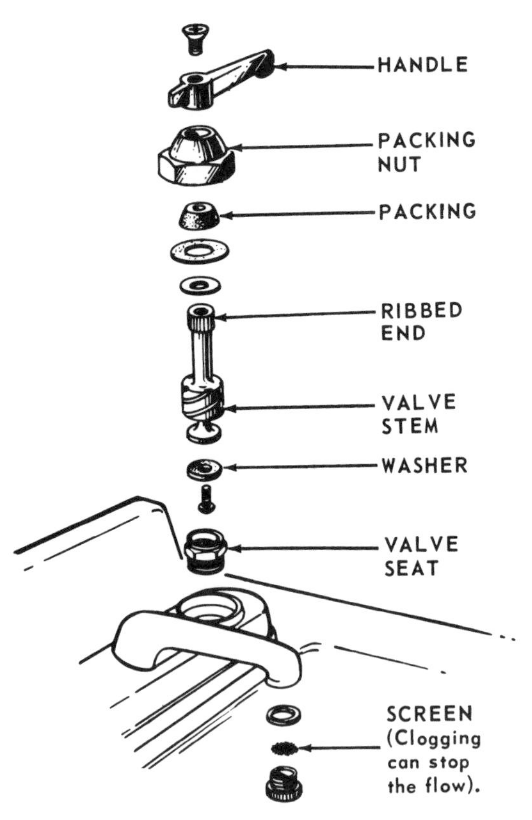

Parts of a typical compression-type faucet.

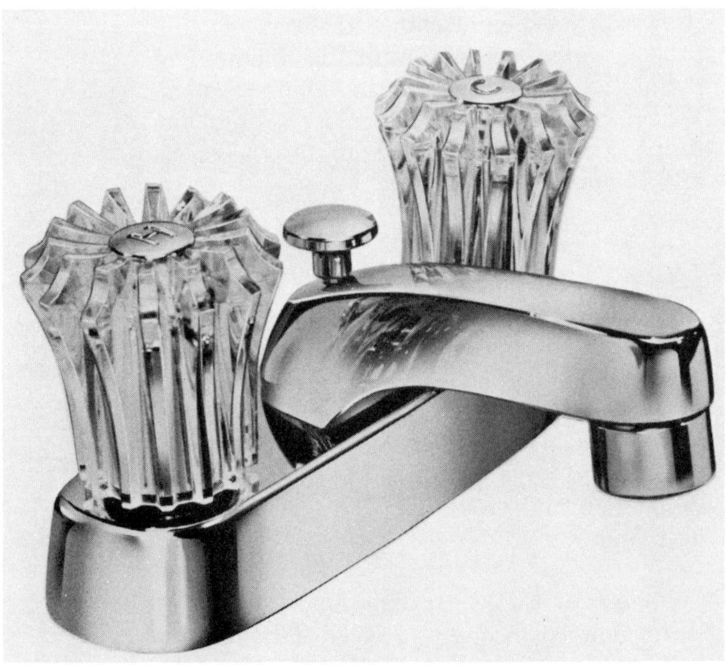

Compression faucets come in all sizes and shapes for just about any purpose. This one is a mixer type for a bathroom sink.

Noncompression faucets feature a single handle or lever to control both the flow and the hot and cold water mix.

against a smoothly finished ring or ground seat surrounding the flow opening, thus shutting off the flow.

If the washer and seat do not make a firm contact at all points, the faucet will leak. This can usually be blamed on a worn washer, which is easily replaced. Most bathtub and kitchen sink faucets have removable seats that can be replaced when they are worn. Seats that are not removable can be reground when necessary.

Noncompression Faucet: This type of faucet has a single lever or knob that mixes and controls the flow of both hot and cold water by means of a valve, ball, or cartridge.

Valve noncompression faucets use water pressure and spring pressure to regulate the mix and flow. Ball types control the water mixture and flow by means of a ball within the faucet. Cartridge types regulate the water through several ports that are located in a cartridge at the top of the faucet.

Repairs of all three types of noncompression faucets usually involve replacing part or all of the mixture and flow control.

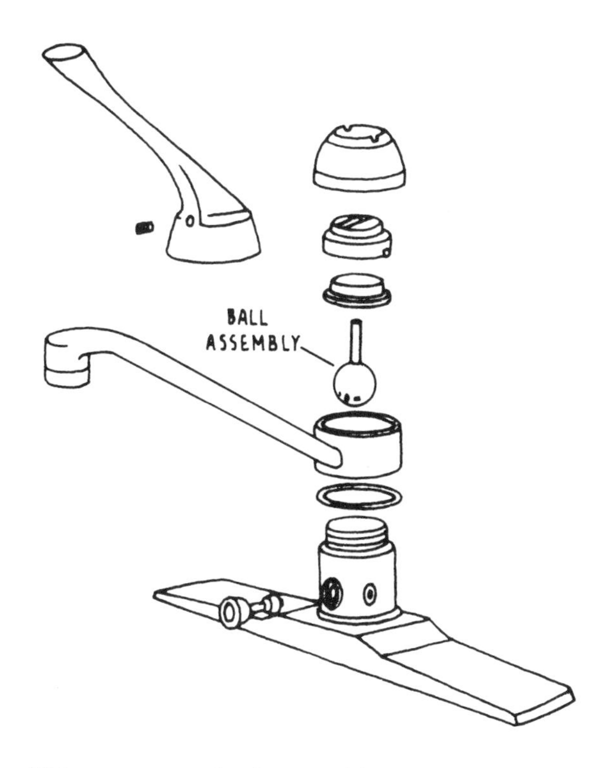

This noncompression faucet utilizes a ball-type control.

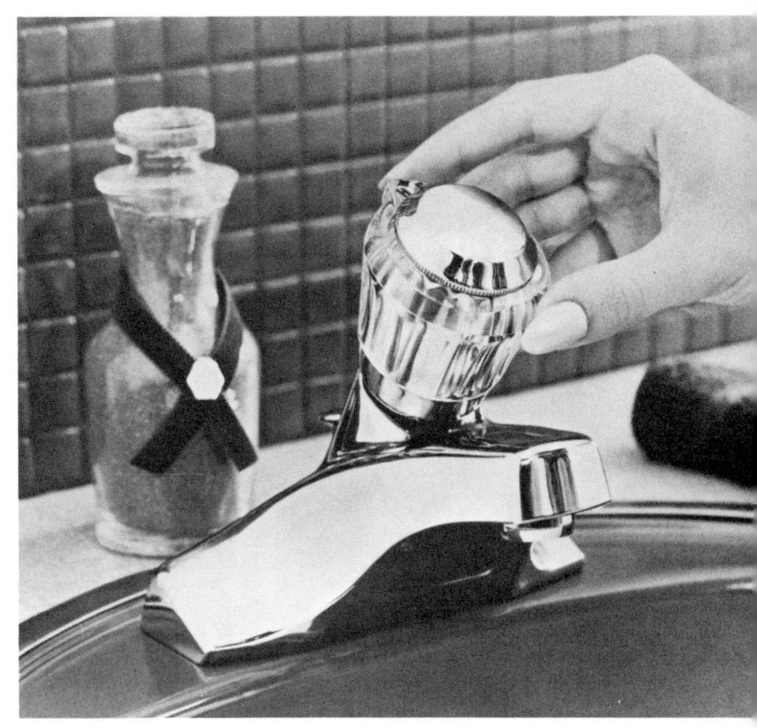

Above and below: *This type of noncompression faucet has a cartridge-type control regulating water flow through ports in the faucet body.*

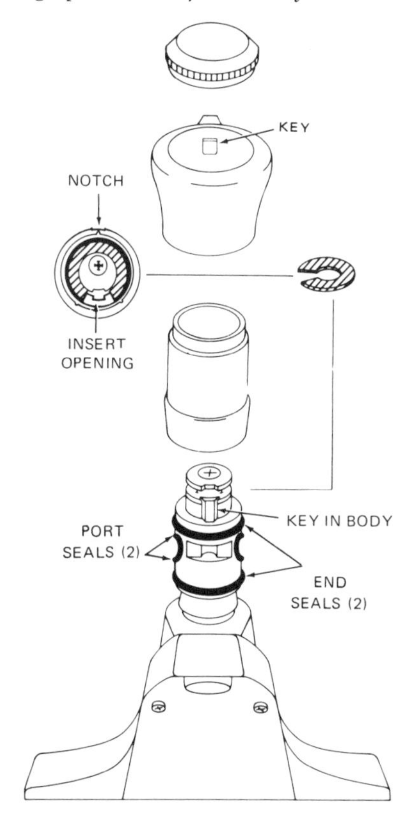

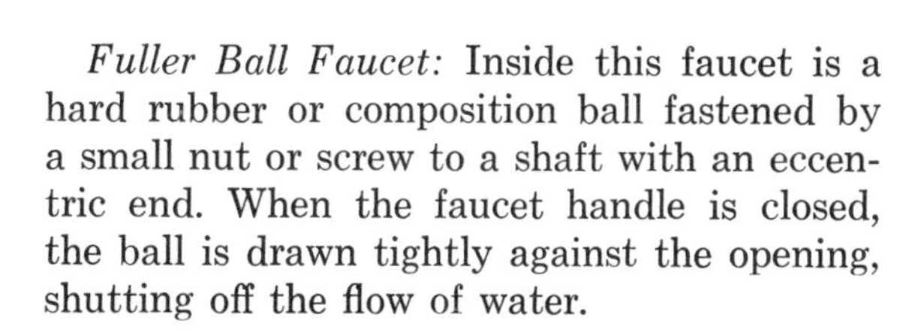

Fuller Ball Faucet: Inside this faucet is a hard rubber or composition ball fastened by a small nut or screw to a shaft with an eccentric end. When the faucet handle is closed, the ball is drawn tightly against the opening, shutting off the flow of water.

Ground-Key Faucet: This type of faucet has a tapered cylindrical brass plunger or plug that fits snugly into a sleeve bored vertically through the body of the faucet. The plunger has a hole or slot bored horizontally through it, corresponding with a similarly shaped opening in the body of the faucet. The plunger is turned by a handle; when this handle is parallel to the body of the faucet, the openings are in line and water passes through.

Faucet Spray: This accessory works by means of a diverter or butterfly valve located inside the faucet in a special chamber. This

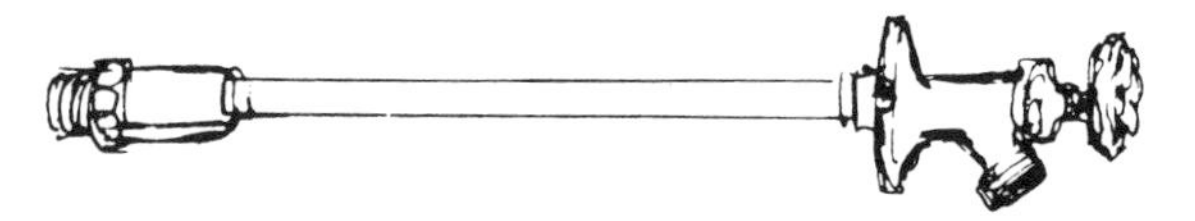

A hose faucet, sometimes called a frostproof faucet, is located on the outside of the house, but its workings are inside where it's warm. This eliminates the need for shutting off the water in cold weather.

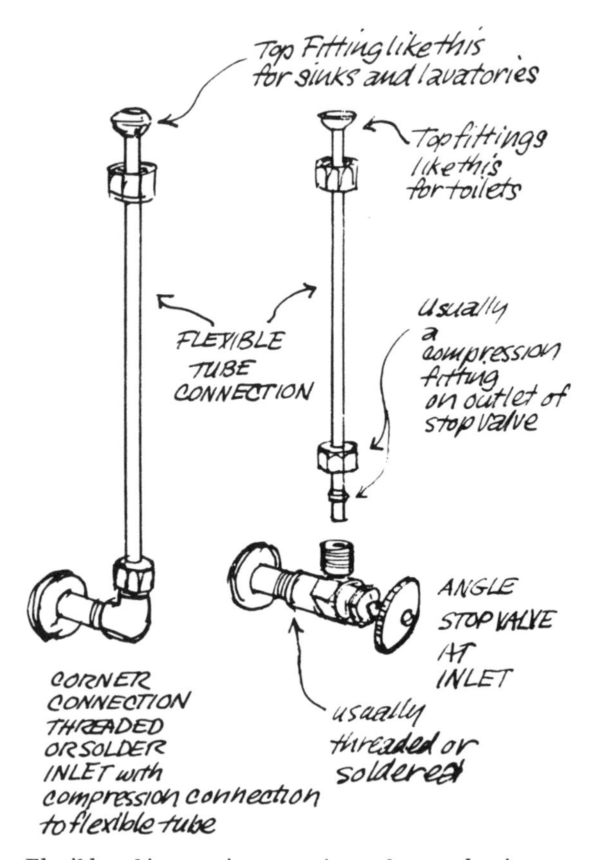

Flexible tubing carries water from the supply pipes in the wall to individual fixtures.

valve normally permits water to flow freely through the faucet spout when the spray attachment is not in use. However, when the thumb-controlled valve on the spray handle is opened, an imbalance of water pressure is created, operating the butterfly valve. A pistonlike unit snaps down, shutting off most of the water supply through the spout and diverting it through the hose that leads to the spray head.

A faucet valve cannot be installed in a faucet that was not originally designed with a chamber to house it.

Hose Faucet: Located on the outside wall of the house, heavy brass hose faucets look and work like any ordinary faucet. However, water flow valves are actually inside the building. The flow valves are controlled by elongated stems. These frostproof faucets eliminate the need for inside shutoff valves.

Fixture Supply Lines

In many instances—for example, a bathroom sink or a toilet—piping is exposed for a short distance between where the supply line comes out of the wall and where it is attached to the fixture. Chrome-plated fixture supply lines are generally used for best appearance. Available in rigid and flexible types and with appropriate fittings at both ends, supply lines are normally ordered with the fixtures. Replacements can be found at hardware and plumbing supply stores.

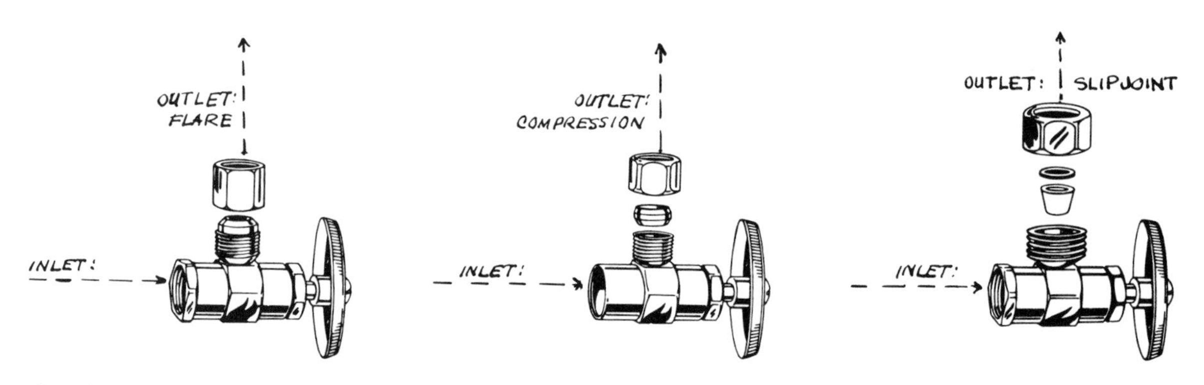

Supply stops are optional but very helpful shutoff valves between water pipes and fixture supply lines.

Pipe Data

Type of pipe	Ease of working	Water flow efficiency factor	Type of fittings needed	Manner usually stocked	Life expectancy	Principal uses	Remarks
Galvanized steel	Has to be threaded. Difficult to cut. Measurements for job must be exact.	Lower than for copper because nipples and unions reduce water flow.	Screw-on connections.	Rigid lengths up to 21 feet. Usually cut to size wanted.	Corrodes in alkaline water more than others. Produces rust stains.	Generally found in older homes.	Recommended if lines are in a location subject to impact.
Copper, hard	Fairly easy to work with.	Highly efficient because of low friction.	Screw-on or solder connections.	12- and 20-foot rigid lengths. (Type-M in 20-foot lengths only.)	Lasts for life of the building.	Generally for commercial construction.	
Copper, soft	Easier to work with than hard copper because it can be bent readily with a bending tool. Measuring a job is not too difficult.	Same as for hard copper.	Solder- or flare-type connections.	30- and 60-foot coils.	Same as for hard copper.	Widely used in residential installations.	
Copper tubing, flexible	Easier to work with than soft copper because it can be bent without a tool. Measuring a job is easy.	Highest efficiency of all metals since there are no nipples, unions, or elbows.	Solder or compression connections.	60-foot coils.	Same as for hard copper.	Widely used in residential installations.	Probably the most popular type used today. Often a smaller diameter will suffice because of low friction coefficient.
Plastic pipe	Can be cut with handsaw or knife.	Same as for copper tubing.	Insert couplings, clamps; also by cement. Threaded and compression fittings can be used.	Rigid, semirigid, and flexible. Coils of 100–400 feet.	Long life. Rust and corrosion proof.	For cold water installations (except CPVC). Used for well casings, septic tank lines, and sprinkler systems. Check local codes before installing.	Lightest of all; weighs about ⅛ of metal pipe. Does not burst in below freezing weather.

4
Assembling Water Supply Lines

Once you have the proper tools—along with pipe and fittings—assembling water supply lines is just a matter of "getting it all together." How this is done depends on the type of pipe you are working with.

Galvanized Steel Pipe

Careful measuring is a most important first step when working with galvanized steel pipe. Allowance must be made for the threads to engage the fittings. The best way to figure this allowance is by the face-to-face method. First measure the exact distance from face-to-face of the fittings. Next, add the extra length necessary for screwing into the fittings:

Pipe size (inches)	Length of pipe screwed into fitting (inches)
½	½
¾	½
1	⅝
1¼	⅝
1½	⅝
2	¾

Remember that the above lengths must be doubled when you are measuring pipes that will screw into fittings at both ends.

Once you have carefully measured the entire job, you can save yourself a lot of work by ordering the pipe already cut and threaded. If, for some reason, you perform these operations yourself, you will need a vise, pipe cutter, reamer, and stock and dies of the appropriate size or sizes to thread the pipe (see pages 17–18). When threading, apply plenty of cutting oil to both the dies and the pipe. After threads are cut and the dies have been removed, wipe off the surplus oil and clear all of the metal chips away from the threads.

To insure watertight connections, you must use pipe joint compound or a special Teflon tape on all joints. Apply the compound or tape to the male

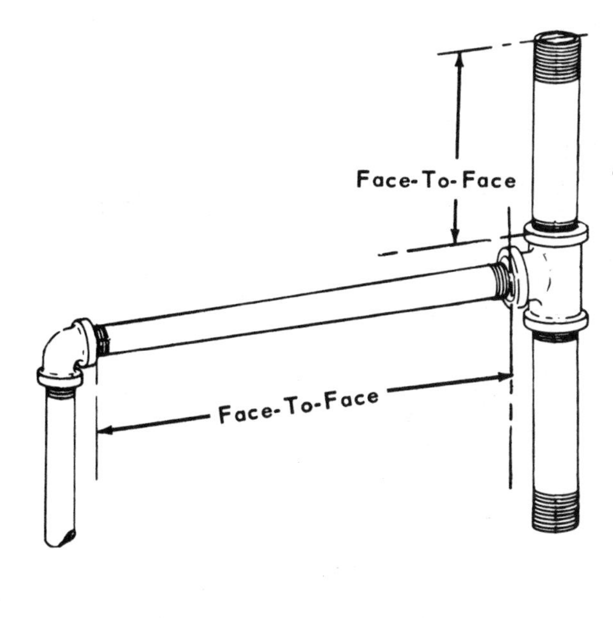

Measuring pipe by the face-to-face method. You must add the distance that the pipe fits into the fittings at each end to the length.

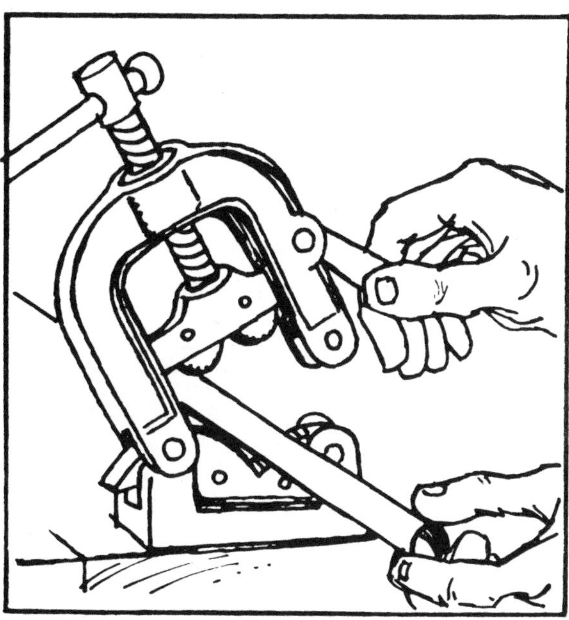

Clamp the steel pipe firmly in a pipe vise.

Cut the pipe with a hacksaw or a pipe cutter (see Chapter 2).

Use a reamer to remove burrs from the inside of the cut end; file burrs from the outside edge.

Start the thread cutter with the dies at a right angle to the pipe. Turn slowly until they take hold.

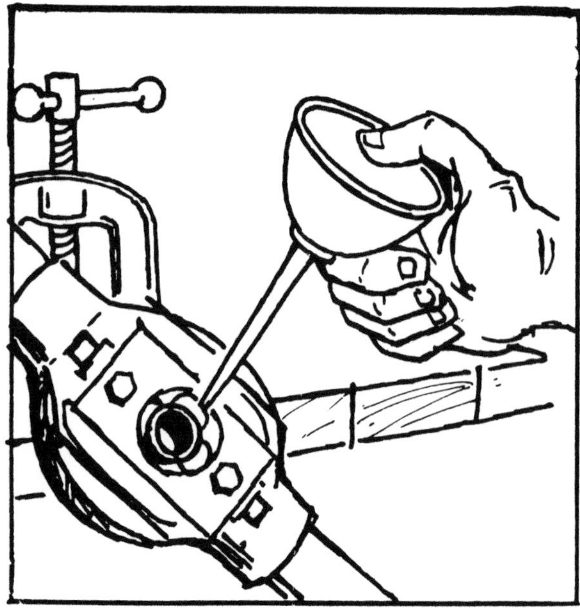

Rotate the cutter a half turn; then back off a quarter turn and apply cutting oil to reduce the friction and heat. Repeat until the thread is cut.

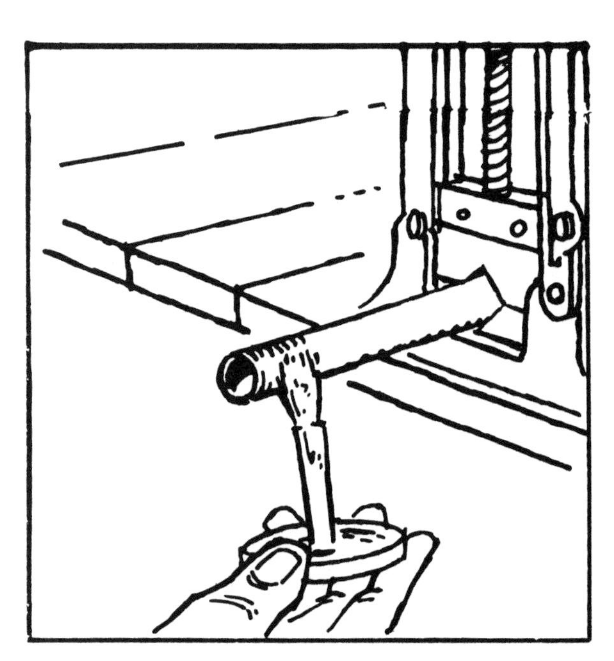

After both ends of the pipe have been threaded, apply pipe joint compound to the threads before adding the fitting.

threads. Fittings may be connected while the pipe is held in a vise. Start the fitting slowly by hand, making sure it is lined up properly and that the threads are engaged. Then tighten it with a pipe wrench. Wipe off the excess compound.

For connecting fittings to pipes that are already in position, two wrenches are needed—one to hold the pipe and one to turn the fitting. Never attempt to tighten a fitting with only a single wrench; you may loosen a connection elsewhere in the line. Again, start the fitting by hand before drawing it up with the wrench.

Rigid Copper Tubing

Rigid copper tubing is measured face-to-face in the same way as galvanized steel pipe. But in this case, the add-on factor is the depth of the soldering hubs in the fittings to be used.

Copper tubing is considerably easier to cut than galvanized pipe. Using a tubing cutter produces neat results. A hacksaw can also be used, but you should first build a simple wooden jig, especially if you will be cutting a lot of tubing. Cut a shallow V-groove along the length of a block of wood (a scrap of 2 × 4 about 2 feet long will do nicely); near one end, cut a slot across the block at a right angle to the groove. Hold the tubing in the groove; the slot will guide the hacksaw, enabling you to get a square, even cut. Use a fine-tooth hacksaw blade (preferably a No. 24). After cutting, remove all burrs by reaming.

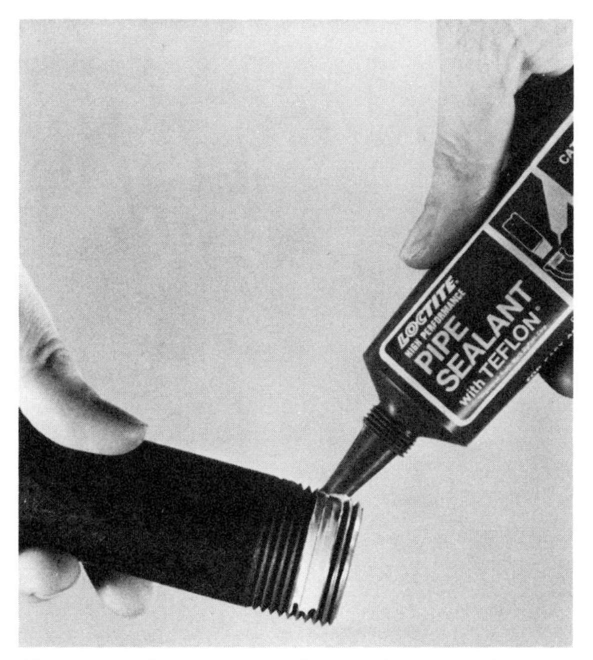

Easy-to-use joint compound comes in tube and even in tape form.

When pipe must be worked where it is already in place, use two pipe wrenches—one acts as a vise; the other turns against it.

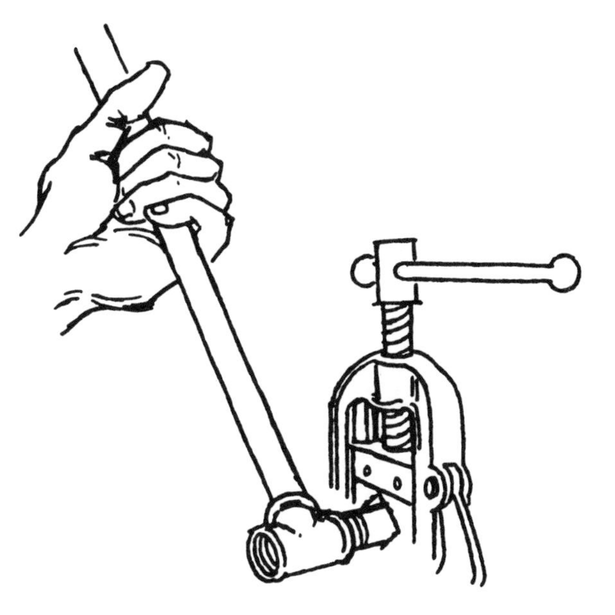

If possible, assemble the pipe and fittings while the pipe is still held in the vise.

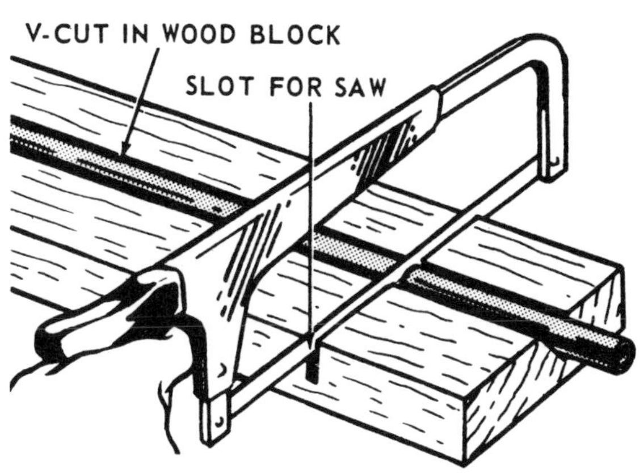

A simple wooden jig is a big help when you are using a hacksaw to cut copper pipe.

Use steel wool or a fine emery cloth to clean and brighten the ends of the tubing and the insides of the fittings to be soldered. Clean surfaces are essential. Make sure that there are no dents in the tubing end, and that it is perfectly round.

Apply a thin coating of noncorrosive flux or soldering paste to the cleaned portions of both the tubing and the fitting. Place the tubing in the fitting up to the hub and rotate it a few times to spread the flux evenly. Be sure to remove excess flux from the outside of the fitting.

No vise is needed when you use a tubing cutter to cut copper pipe. Rotate the cutter around the pipe while tightening the handle.

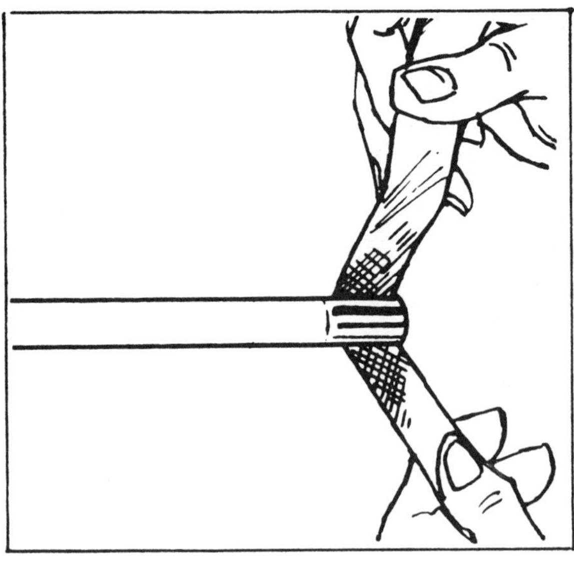

Clean the end of the tubing and the inside of the matching fitting with an emery cloth.

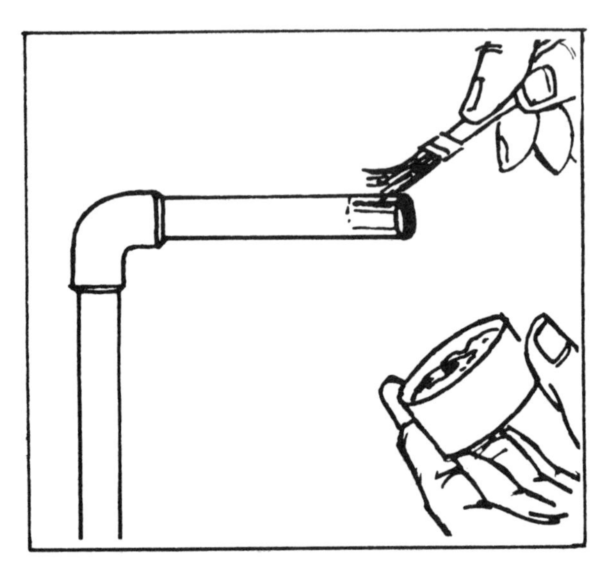

Apply a thin film of flux to the end of the pipe and the inside of the fitting. Place the fitting onto the pipe and rotate to spread the flux evenly.

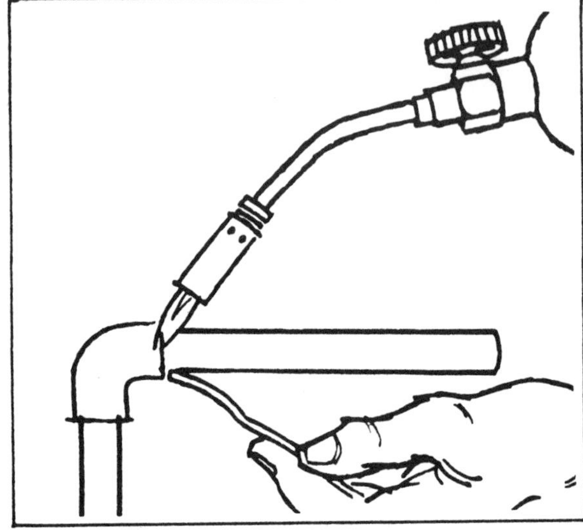

Apply heat with a propane torch and hold solder to the joint until capillary action causes it to flow around the connection.

Apply the flame of a propane torch directly to the fitting, heating it evenly. When the flux begins to bubble out, remove the torch and apply the end of the solder to the edge of the fitting. Capillary action will draw the solder into the space between the tubing and the fitting, filling and sealing the joint. Make sure a line of solder shows completely around the joint, and do not hold the flame on the connection after it is filled. Wipe away excess solder with a rag.

Make sure the pipe and fitting do not move

Assemble the pipe and fittings on the floor or workbench before lifting into position.

while the solder is cooling—this could cause a weak joint or broken seal. The solder should cool and harden in less than a minute.

When other soldered connections must be made to the same fitting (such as with an elbow), wrap the finished joint with wet rags to prevent the solder from melting.

If a soldered joint should leak, or if it is necessary to disconnect it for another reason, the water must be drained from the entire line before repairs begin. Then wrap wet rags around other connections in the same fitting to keep them from melting. Heat the leaking joint until the solder runs and then pull out the tubing. Clean both tubing and fitting with steel wool or an emery cloth before resoldering.

Types of Solder

- "Soft" solder is 95 percent tin and 5 percent antimony. It flows at 465° Fahrenheit and hardens quickly to make a strong joint.
- "50–50" solder is 50 percent lead and 50 percent tin. It flows at 250° Fahrenheit and is slow to harden.
- "Hard" solder is an alloy of copper and phosphorus. It flows at high temperatures —above 1300° Fahrenheit—and is the best for making strong joints.

Flexible Copper Tubing

Flexible copper tubing can be used with either soldered or flared fittings. With soldered fittings, it is handled in exactly the same way as rigid tubing. However, greater care must be taken when cutting so that the ends are not dented or flattened.

When measuring pipe for use with flared fittings, use the face-to-face method. Add approximately $\frac{3}{16}$ inch for each fitting. If the end is slightly flattened during cutting, flaring will round it out. After cutting, file off any burrs (the end of the tubing must be perfectly smooth).

Flare-type fittings have threaded flange nuts to hold the tubing in place. Remove the flange nut from the fitting and slide it onto the tubing before flaring the tube end.

Loosen the die block clamp screw of the flaring tool and insert the tubing in the cor-

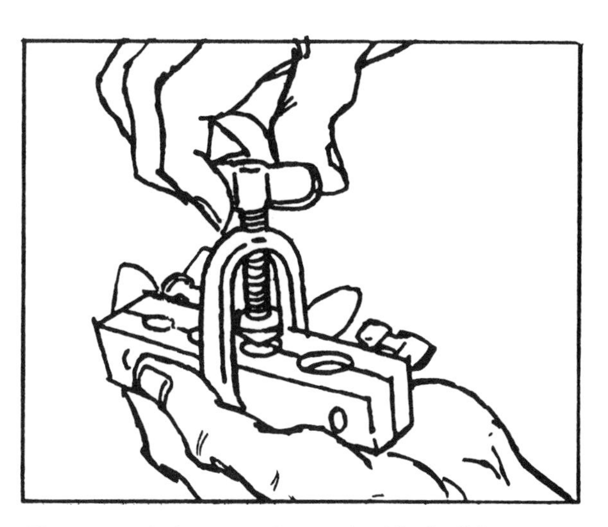

Flare-type fittings can be used with flexible copper tubing. After cutting and reaming, slip on fitting nut. Clamp tubing in the die block and tighten down the flaring tool handle.

responding size hole. The tubing should extend approximately $\frac{1}{16}$ to $\frac{1}{8}$ inch above the face of the block; tighten the clamp screw to hold the tubing firmly. Slide the yoke over the end of the die block and turn the feed screw clockwise until the flaring cone forces the end of the tubing tightly against the grooves of the die block, forming a bell-shaped flare. Back out the feed screw, slide the yoke off the die block, and loosen the clamp screw to remove the flared tube.

Slide the flange nut up to the end of the tubing (it can't come off now because of the flare). Start the nut on the threads of the fitting, hand-tightening it to make sure it lines up correctly. Finish making the connection by tightening the nut with two wrenches (open-end or adjustable open-end). Use one wrench on the nut and the other on the fitting. This forces the flare at the end of the tube to come against the cupped end of the fitting, forming a tight seal. As with galvanized pipe and fittings, never use only one wrench when tightening.

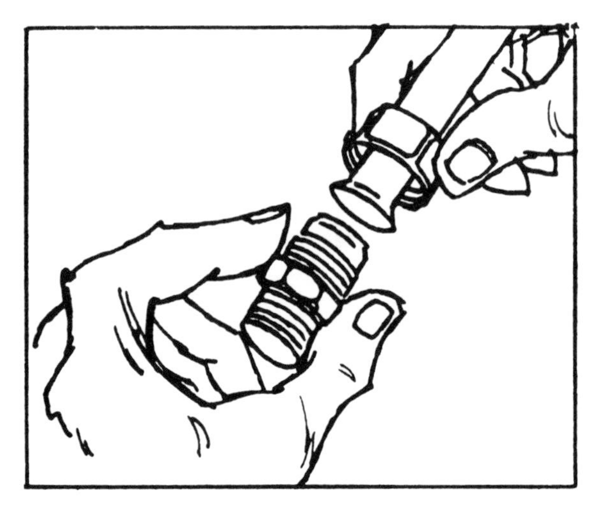

Place the flared end of tubing against the matching cone of the fitting; slide the nut onto fitting and tighten securely.

Plastic Pipe

Plastic pipe is easiest for the do-it-yourselfer to install. Once again, use the face-to-face method to measure the pipe, adding the length that the pipe will be inserted into the fittings (with most fittings, the length to be added is the same as the diameter of the pipe).

Flexible polyethylene tubing can be cut with a sharp knife. Semirigid and rigid plastic tubing is cut with a hacksaw or other hand saw—or even a power saw. Use a fine-toothed blade (9 to 14 teeth per inch) with little or no set. Never use a rotary pipe cutter on plastic tubing.

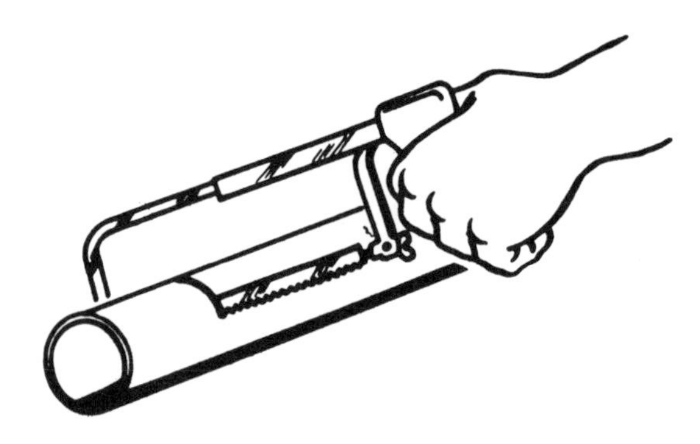

Use a hacksaw, hand saw, or power saw to cut plastic pipe. Some flexible types can be cut with a sharp knife.

You can build a simple jig—the same as described for cutting rigid copper tubing—to insure square cuts when using a hand saw. If you place the plastic pipe in a vise, wrap it in cloth or tape to prevent surface damage. After cutting, ream the tubing with a pocketknife.

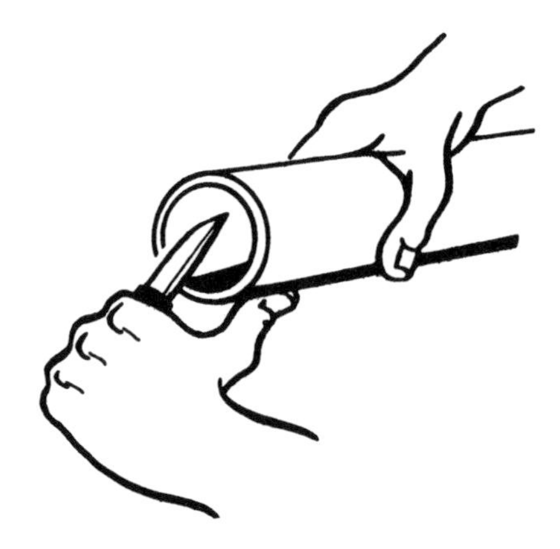

Ream the pipe using a standard reamer or a pocketknife.

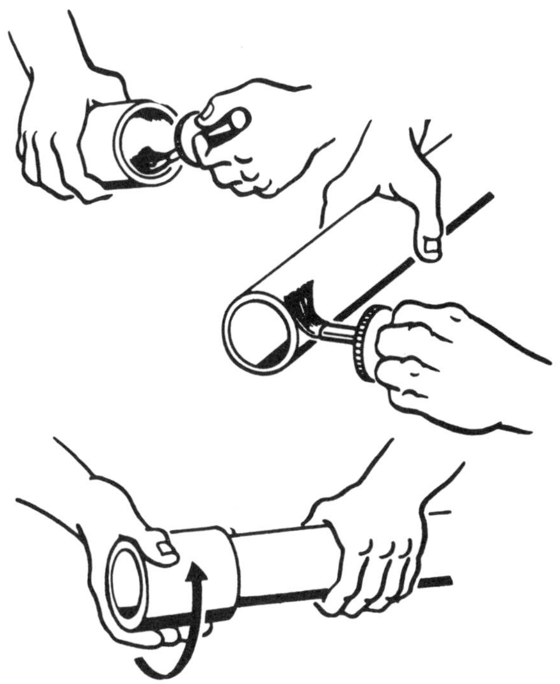

After cleaning, apply solvent to the inside of the fitting and the outside of the pipe where they will join; then place the fitting on the pipe and turn slightly to spread the solvent evenly. Hold them together until the solvent begins to cure.

The most common means of connecting plastic pipe and fittings uses a special solvent. Use an emery cloth to clean both the inside of the fitting socket and the outside of the pipe end (at least as far as the pipe will fit into the socket). Make sure to get the proper solvent for the type of plastic pipe you are using. Simply buy your solvent when you purchase the plastic pipe (see Chapter 3, Plastic Pipe). With a nonsynthetic bristle brush, apply the solvent generously to the inside of the fitting socket and to the outside of the pipe. Press the fitting firmly onto the pipe and give the pipe a quarter turn to distribute the solvent evenly. Make sure that the pipe and fitting are properly aligned. Hold them together tightly for about 15 seconds so that the pipe does not push out of the fitting until the curing process has begun. Clean off the excess solvent. Make sure the joint is not moved until the solvent weld has set. Although the solvent sets quickly, allow at least 12 hours before testing the line under pressure.

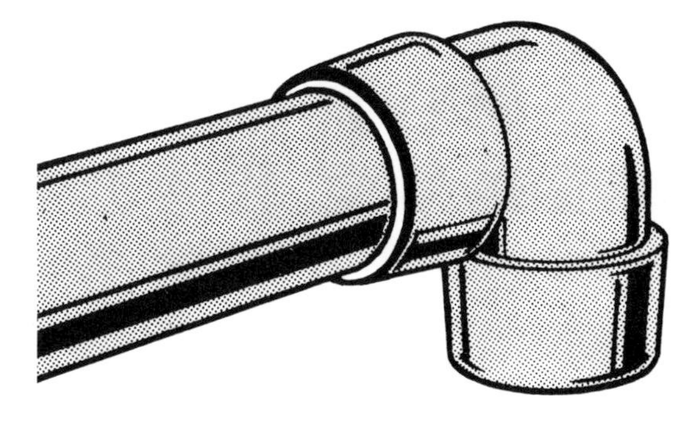

Excess solvent can be wiped away, but make sure there is a continuous bead around the joint.

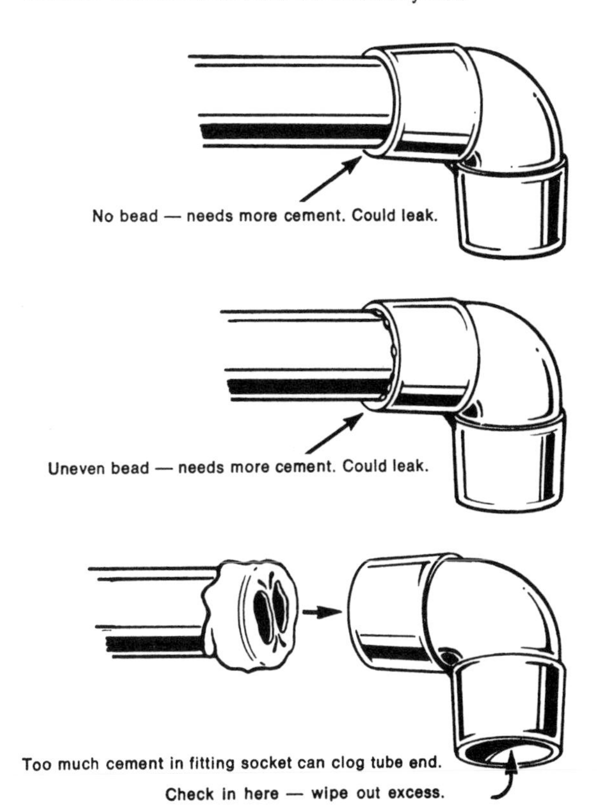

Too little solvent could mean a leaky joint; too much can clog the pipe and prevent the flow of water. With a little practice, you will be able to make perfect joints.

Once set, a solvent-welded joint is permanent. If the joint has to be broken, the fitting must be sawed off and a new fitting installed in its place. If necessary, you may be able to salvage the fitting by sawing through the pipe a short length away on each side and then reassembling the line with couplings.

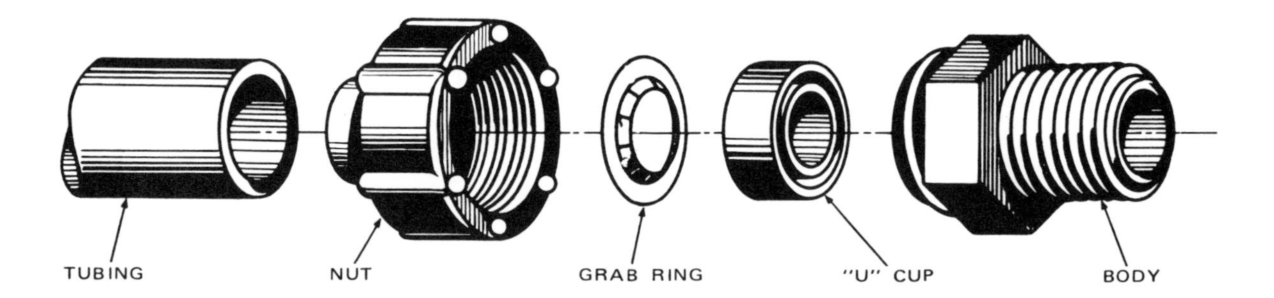

Above and below: *Unique fitting requires no solvent to make a watertight joint. The tubing is pushed into the fitting and held fast by a special U-cup. Water pressure inside the tubing anchors it in place. The fittings can also be used on copper tubing.*

There is another type of plastic fitting—one that doesn't require solvent. The pipe end is cleaned and smoothed with an emery cloth (as for solvent connections). The pipe end is then pushed into the fitting, which has a special nut. The nut is hand-tightened to make a watertight joint. What makes the fitting work is a unique collar—called a *U-cup*—through which the tubing is inserted. Water presses one side of the cup against the tubing and the other side against the body of the fitting—the higher the water pressure, the tighter the seal.

5
Drainage Lines

The average household uses about 200 gallons of water each day. The drainage system must be capable of getting rid of this water quickly and efficiently. Before making any additions or alterations to your home's drainage system, be sure you check local plumbing codes.

Cast-Iron Drainpipes

Cast iron is one of the strongest and most durable pipe materials made (the fountains at the Palace of Versailles have been operating for over 300 years with their original cast-iron pipes). It is the type most generally used for soil stacks.

Cast-iron pipe comes in 5- and 10-foot lengths. The traditional type has a "bell" or hub at one end for connecting to fittings and other lengths of pipe.

Drainage pipes roughed-in for back-to-back bathrooms. The lavatory traps are shown already in place (left); in actual practice, they would be installed later. The toilet drains (center) empty directly into main stack.

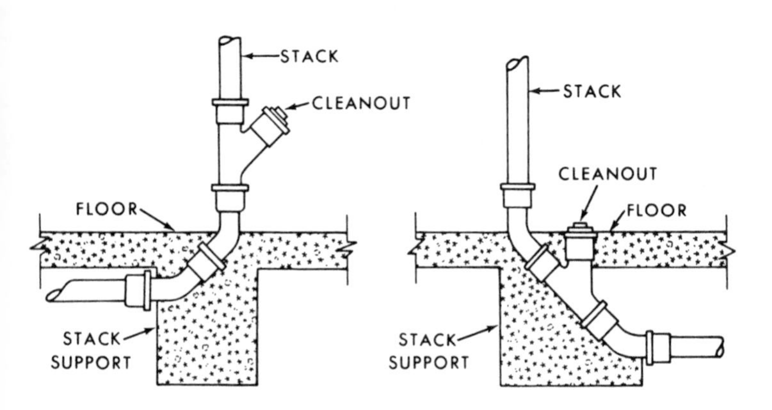

A cleanout plug is provided somewhere near the bottom of the main stack. Note that concrete is poured beneath the floor to support the heavy stack.

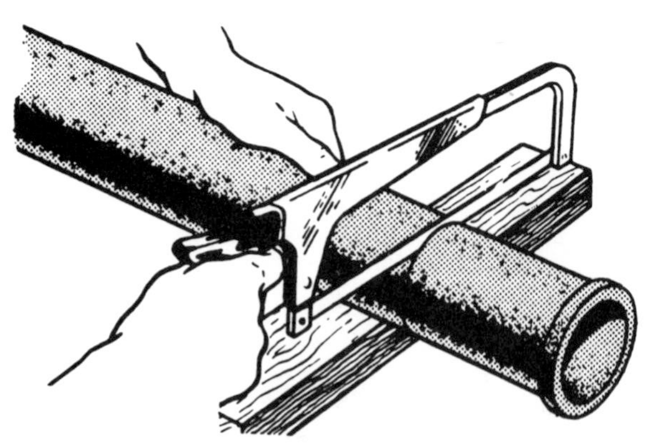

Mark a chalk line around the pipe where it is to be cut; then make a shallow cut all around with a hacksaw.

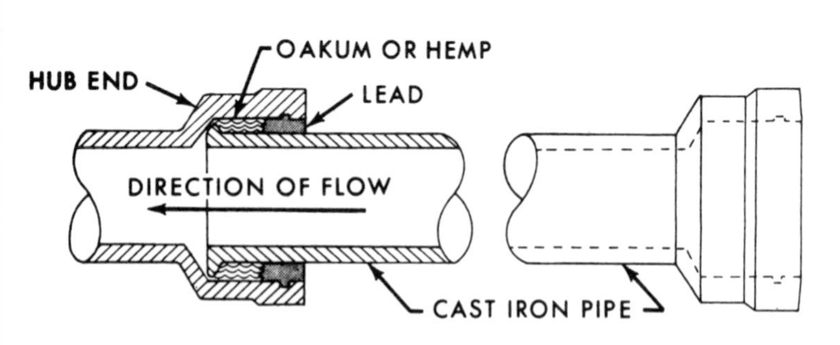

A bell-and-spigot joint in cast-iron pipe.

With a hammer, tap the pipe all around the score until it breaks off.

A fairly new type of cast-iron pipe (hubless) eliminates this bell; both ends are plain, and connections are made by slipping the pipe ends into a neoprene shield. Then clamps are used to tighten the shield over the pipes and fittings.

When measuring for hubless pipe, measure tight; that is, the ends of joining pipes or of pipes and fittings butt together. When measuring standard cast-iron pipe, allow additional length for fitting into the hub. Allow 2½ inches for 3-inch pipe and 3 inches for 4-inch pipe. When pieces of pipe less than 5 feet long are needed, use double-hub pipe, which comes with a hub on each end. When the pipe is cut, each end has a hub; otherwise, the cut-off end would be useless.

With a piece of chalk, mark around the pipe where it is to be cut. Use a hacksaw to make a $\frac{1}{16}$-inch-deep cut all around the pipe, keep-

Extra-heavy pipe will have to be chiseled around the scored line until it breaks off.

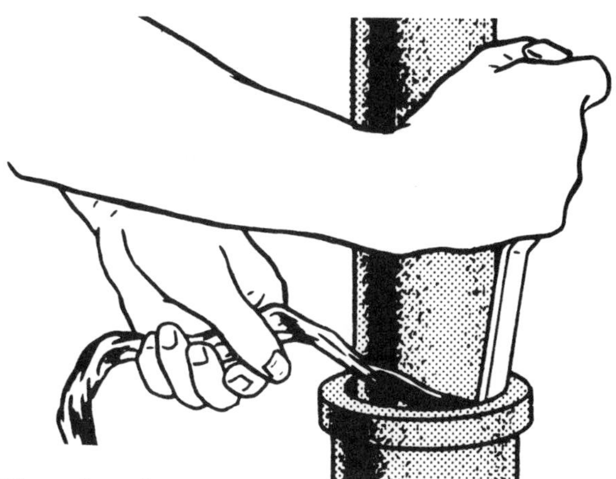

Place the spigot end of the upper pipe into the hub of the lower pipe. With a yarning iron, pack oakum into the joint.

Make sure the joint is perfectly dry; then carefully pour lead around the pipe until the hub is full.

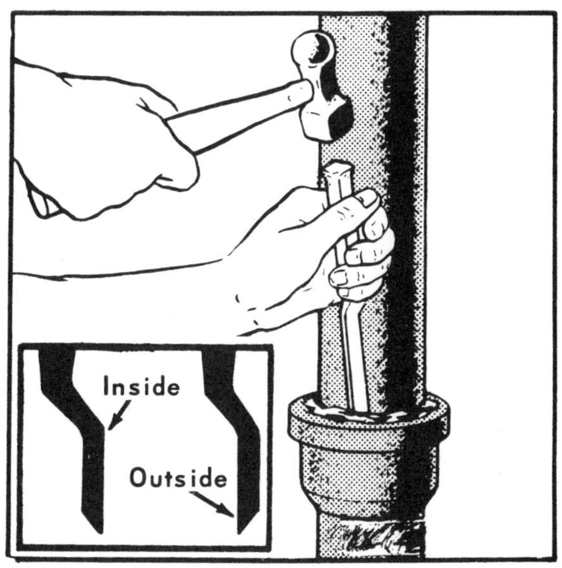

After the lead has cooled, tamp it down firmly, first with an outside calking iron, then with an inside calking iron.

When pouring a horizontal leaded joint, an asbestos joint runner is used to keep the lead from running out of the hub.

ing it square with the pipe to insure a clean, even break. If you are using lightweight pipe, tap around the pipe with a hammer until it breaks at the cut.

Heavy-duty pipe is more difficult to break. After making the hacksaw cut, place the pipe across a piece of 2 × 4 laid flat on the floor. Using a cold chisel and hammer, strike lightly all around the scored line, holding and turning the pipe with your knee. Continue striking around the pipe, hitting the chisel harder each time, until the pipe breaks off.

With hubless pipe, you simply add the sleeve to make the connection. With standard cast-iron pipe, leaded joints must be used.

Leaded Joints

Vertical Joints: When erecting a vertical cast-iron drain line, position each piece with the hub end up. Make sure the ends of the pieces to be joined are clean and dry. (Moisture causes molten lead to fly out of the joint, which could result in serious injury. *Wear goggles and gloves when pouring lead —and keep out of range even if the joint appears dry.*) Place the spigot (plain) end of the next higher section into the hub to its full depth. Temporarily brace it in position. Check with a level or plumb bob to make certain that the two sections to be joined are perfectly straight up and down. Also make certain that the spigot end is centered in the bell.

For small pours, a blowtorch can be used to melt lead in the melting pot. If the line requires several pours, a plumber's furnace

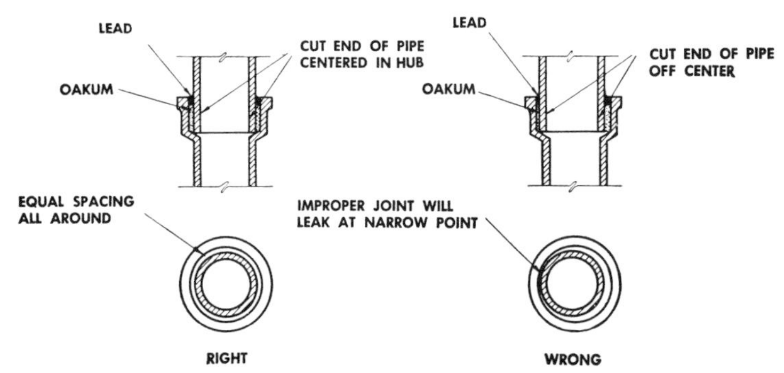

Always make sure that the spigot of one pipe is centered in the hub of the pipe it is being joined to. Otherwise, you will have a faulty joint.

should be used. Light the furnace and place the calking lead in the pot. If your local code requires one inch of lead in the joint, use a pound of lead for each inch of pipe diameter (that is, use 4 pounds of lead for 4-inch pipe, 3 pounds of lead for 3-inch pipe, and 2 pounds of lead for 2-inch pipe). If your local code requires ¾ inch of lead in the joint, use 3 pounds for 4-inch pipe 2¼ pounds for 3-inch pipe, and 1½ pounds for 2-inch pipe.

While the lead is melting, pack the joint with oakum, which comes in the form of a rope. For 4-inch pipe, you will need about 5 feet of oakum; for 3-inch pipe, 4½ feet; for 2-inch pipe, 3 feet. Wrap the oakum around the pipe at the joint and drive it to the bottom of the hub with a yarning iron. Compress it firmly to make a solid bed for the lead and to prevent leakage into the pipe. Continue packing until the joint is filled to within 1 inch or ¾ inch of the top (depending on the thickness of lead as required by the code).

Never dip a cold ladle into molten lead—this could cause an explosion. Heat the ladle first by placing it alongside the melting pot. When the lead is molten, dip it out and pour it evenly into the hub around the joint. Continue in a single pour until the lead is even with the top of the hub.

When the lead cools, move an outside calking iron slowly around the joint. Tap the joint lightly with a hammer to pack the lead against the hub. Repeat this step with an inside calking iron to pack the lead against the pipe. Tamp the lead firmly all around to make an airtight and watertight seal.

Horizontal Joints: The procedure for making a horizontal leaded joint is similar to the procedure for making a vertical joint. However, an asbestos joint runner must be used when making horizontal joints. The joint runner prevents the lead from running out of the hub as it is poured. Center the spigot in the hub and pack in oakum (see Vertical Joints, above). Place the joint runner around the pipe, fitting it as tightly as possible just above the hub, with the clamp at the top of the pipe to form a funnel for pouring. Tap the runner against the top of the hub to prevent the lead from running out.

Using a full ladle of lead, make the joint with only one pouring. Pour it into the hub until it overflows. Allow the lead to cool and solidify; then remove the joint runner. Cut off surplus lead with a cold chisel and hammer; then calk with outside and inside irons, as with vertical joints.

Galvanized Steel Pipes

Threaded galvanized steel pipes in 1½- and 2-inch sizes are generally used with cast-iron pipe for branch drains, vent lines, and sometimes for secondary stacks. However, steel pipe must never be buried underground.

Galvanized steel drainage pipe is cut, threaded, and assembled in the same way as galvanized water pipe (see Chapter 4).

Copper Drainpipes

Copper drainpipes have thinner walls than copper water supply pipes, since they do not have to withstand pressures. In all other respects they are similar. Copper drainpipes are cut and soldered the same way as copper water supply pipes (see Chapter 4).

Plastic Drainpipes

Plastic drainpipes (such as PVC or polyvinyl chloride) are freezeproof, rustproof, and

Joints in copper drainpipes are soldered following the same techniques used for copper water supply pipes.

corrosionproof. They resist acids and alkalies and are economical, lightweight, and easy to install—even for the most unskilled do-it-yourselfer. However, many local plumbing codes do not yet allow plastic drainpipes, so make sure you check before planning to use plastic pipe for your drainage system.

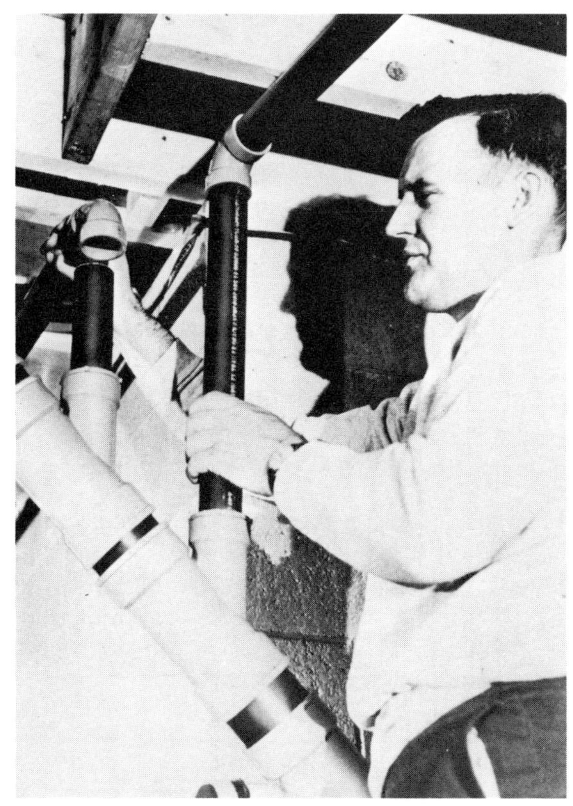

Joints in plastic drainpipe are made quickly with special solvent.

Plastic pipe is usually stocked in 10-foot lengths. It is easily cut with a hand saw, and connections are made with a special quick-setting cement (see Plastic Pipe, Chapter 4).

Drainage Fittings

As with pipe, many fittings are available that let you run your drainage line just about anywhere it has to go. In fact, there are even more configurations for drainage fittings than for pipe fittings. In addition to Ts, there are Ys and Ps (traps), and even Y branches. There are all kinds of reducers and offsets to

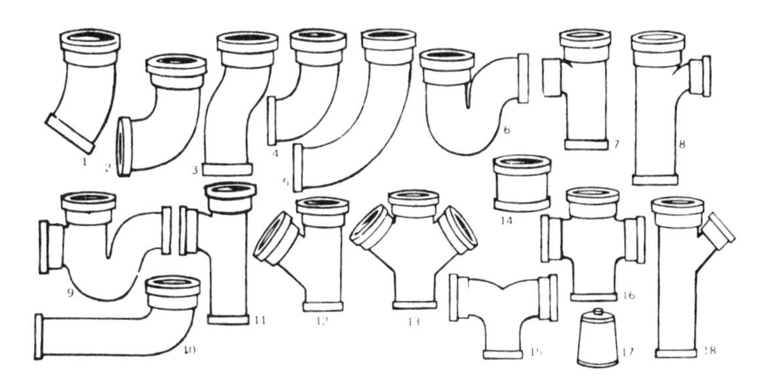

Drainage fittings
1. 1/8 bend
2. 1/4 bend
3. offset
4. reducing bend
5. long sweep
6. P trap
7. short sanitary tee
8. reducing tee
9. running trap
10. long bend
11. long sanitary tee
12. Y branch
13. double Y branch
14. sleeve
15. double 1/4 bend
16. sanitary cross
17. cleanout plug
18. reducing Y branch

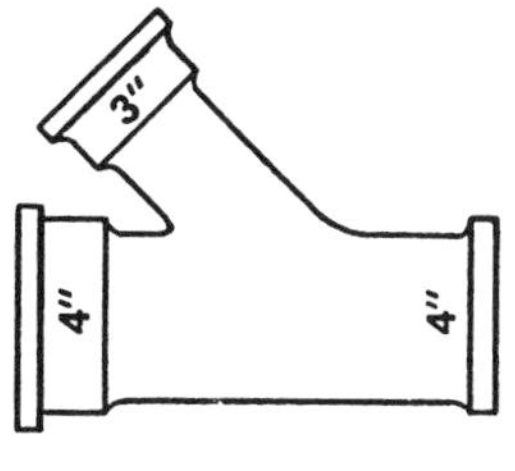

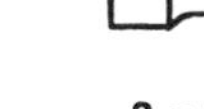

How reducing fittings are measured: at left, a 4 × 3 Y (the 3-inch line is a branch of the 4-inch main); at right, a 3 × 2½ × 2 tee (the 2-inch line joins the others at a right angle).

help you out of tight spots. There are long sweeps and long bends (the long bends are specially designed for connecting to toilets). Elbows are often called *bends* (a ⅛ bend is a 45° elbow; a ¼ bend is a 90° elbow).

Fixture Drainage Lines

Like fixture supply lines, exposed fixture drains are chrome-plated for appearance. Normally they are manufactured of brass.

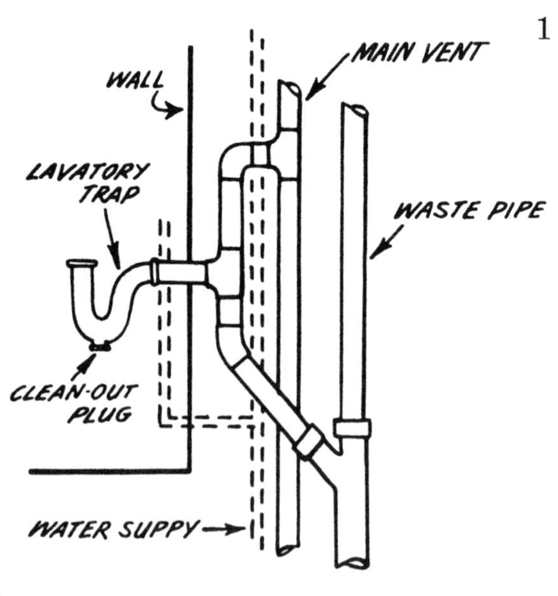

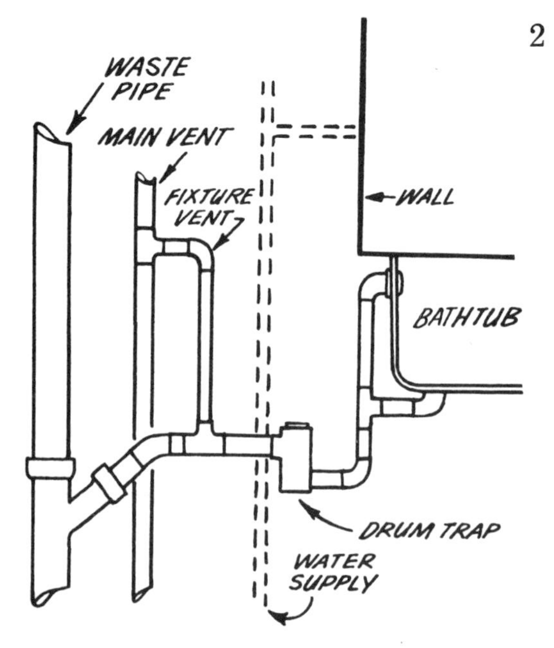

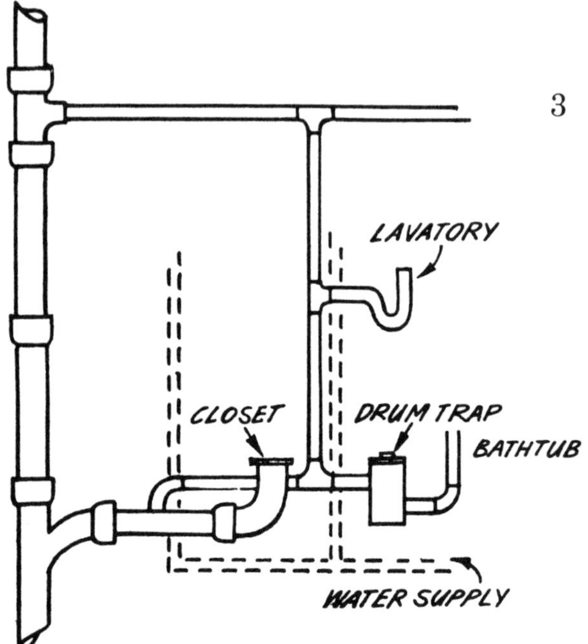

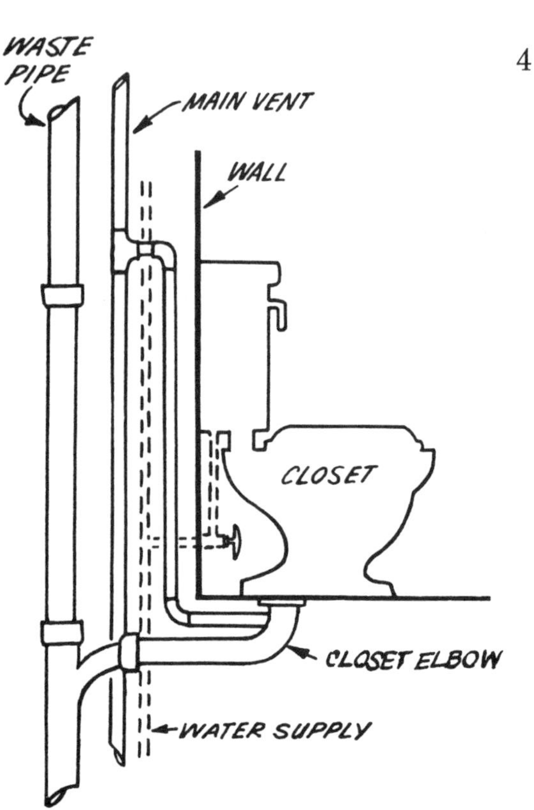

Examples of drainage piping in typical installations. The lavatory installation in number 1 shows the drain line connected to the waste pipe at an angle. Number 2 shows a method of connecting a fixture vent to the main vent. In number 3, all three bathroom fixtures are connected to a common vent pipe. The toilet installation shown in number 4 connects the vent to the main vent above the fixture.

6
New Fittings for Old

Except for some very inexpensive ones (which you should avoid like the plumber's plague), fittings such as faucets and toilet valves are built to take a beating—which they do. Consider how many times a day you turn on and off the faucet in your kitchen sink, or consider how often you flush the toilet. Over the years that's a lot of use. Parts (especially moving parts) wear and are repaired. But eventually they reach a point of no repair. Replacement is the only answer.

Installing a new faucet is a quick way to add convenience and beauty to your kitchen.

There may be other reasons for replacing old fittings as well. Perhaps you are remodeling your bathroom. New tile, fresh paint on the walls, and plush carpeting on the floor can make those old faucets look like they belong in a barnyard bathtub. Shiny new faucets and a modern shower head definitely should be part of your remodeling scheme.

When selecting new fittings, it is best to take the old ones along with you. This will assure your getting the correct replacement size. If this is impractical, measure the old installation carefully; connecting pipe sizes and accurate center-to-center distances must be taken for combination faucets. The method of installing new fittings depends on the manufacturer; the instructions packed with the fittings should be followed to the letter. Some general guidelines are given here. Read them through before you gather your tools and begin work.

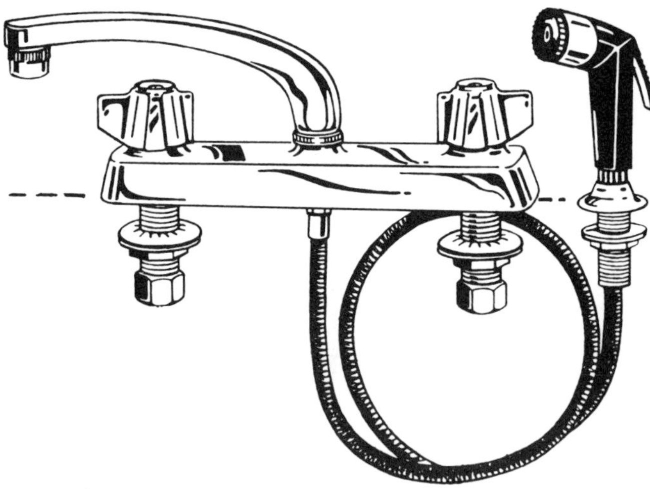

A typical exposed-deck kitchen faucet.

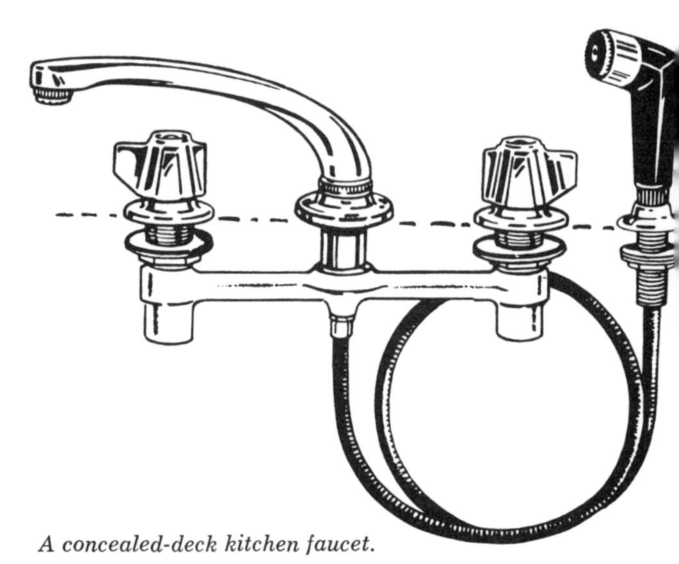

A concealed-deck kitchen faucet.

Kitchen Sink Faucets

Most plumbing systems include stop valves (supply stops) in the hot and cold water lines near each faucet. The first step is to turn these off. If you don't have or can't find these valves, turn off the closest valve you can find or, if necessary, turn off the main valve (shutting off all water in the house). Open the faucet to drain out as much water as you can.

When installing a typical mixer-type sink faucet, first remove the nuts that hold the supply line to the old fitting; then remove the hexagonal nuts that hold the fitting to the sink. In most cases, a basin wrench is the best tool for this job. If the fitting includes a spray hose, remove the nut holding the hose to the faucet. The old faucet then lifts out.

Wipe clean the top of the sink where the new faucet will be installed. Most exposed-deck faucets (the deck is the part that connects the various components of the faucet) have a built-in rubber gasket on the bottom; if your new one does not, put a ring of plumber's putty on the sink around the perimeter where it will be placed. Remove the spray hose (if it has one) from the new fitting and set the faucet in place. Tighten the nuts on the faucet shanks and then reconnect the supply lines. Slip the spray hose through the guide sleeve in the sink and attach it to the faucet, completing the installation.

In some older sinks, the faucets are placed in the backsplash against the wall. Often, the supply lines come directly out of the wall. To replace these, first shut off the water at the nearest valve (probably in the basement). Then simply remove the slip nuts located behind the backsplash that hold the faucet shanks in place. Remove the old faucet and put the new faucet in place.

Single faucets are seldom seen in kitchen sinks these days. They usually can be re-

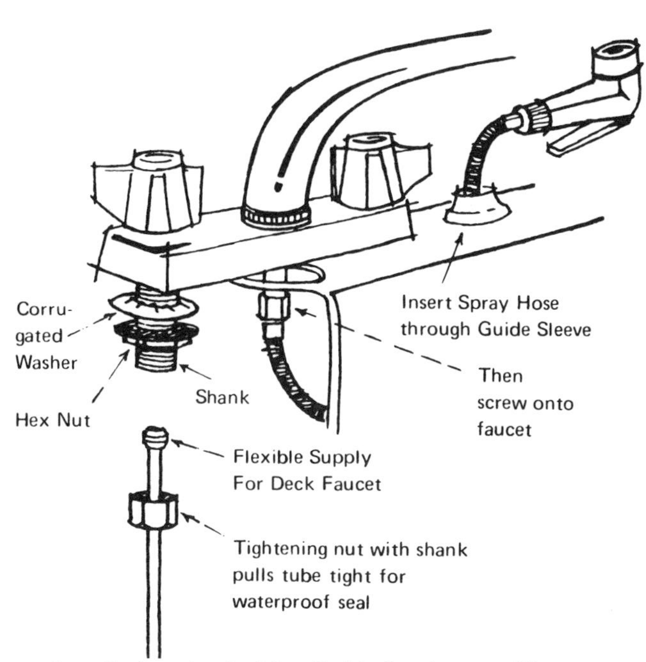

Installation of a dual-handle kitchen faucet with spray hose attachment.

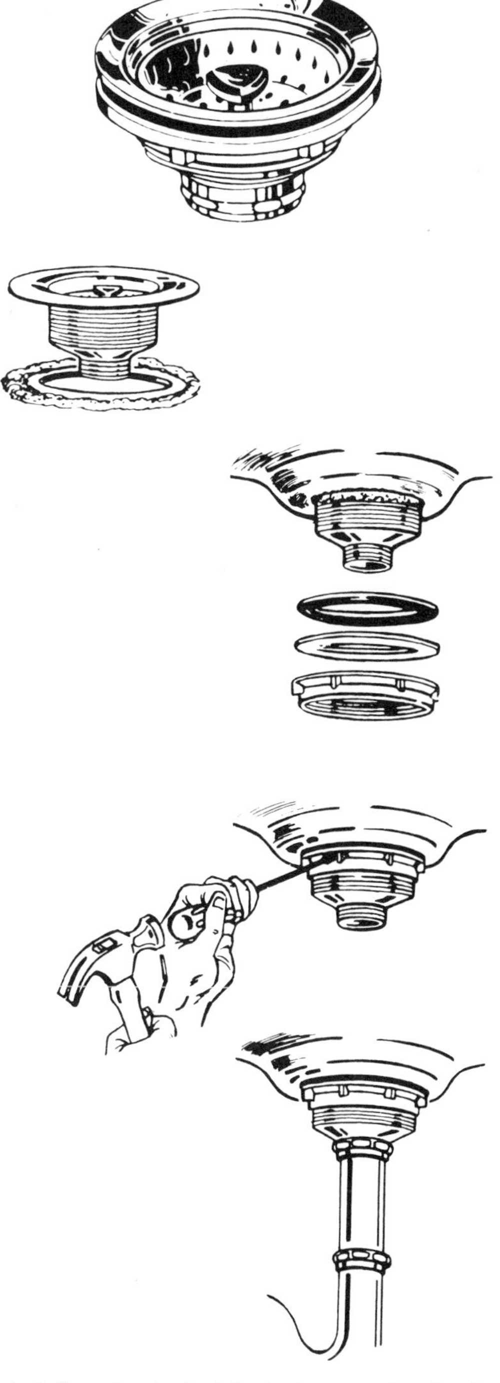

placed with newer ones simply by unscrewing them from the supply lines to which they are directly connected. Replacement single faucets come with the inlet supplied with either male (outside) or female (inside) threads, so check your present installation before buying a new one.

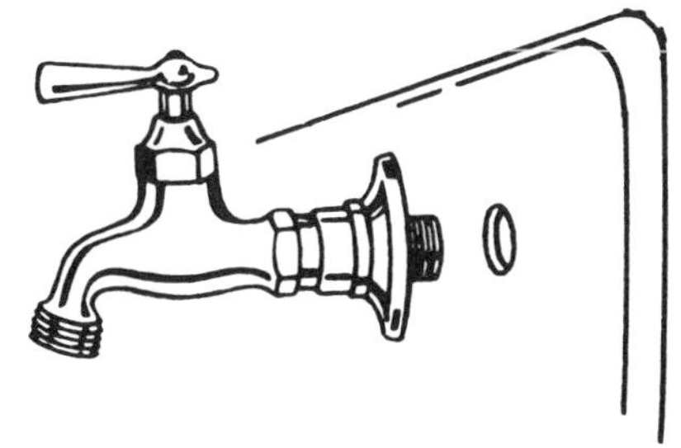

Single-sink faucets are supplied with either male or female threads. Some also have a hose attachment.

Sink Strainers

Kitchen sink strainers (the built-in kind), which trap waste material before it enters drainpipes and double as plugs to hold water in the sink, sometimes need replacement—especially if the open–close mechanism becomes damaged.

To install a standard sink strainer, apply plumber's putty around the opening; then remove the nuts and washers from the strainer and place them through the opening. From beneath the sink, place the washers over the strainer threads and tighten on large metal nut. While an assistant holds the strainer from above, complete tightening by tapping with a screwdriver and hammer.

Above and below: *A sleek, single-handle, cartridge-type lavatory faucet with a pop-up drain.*

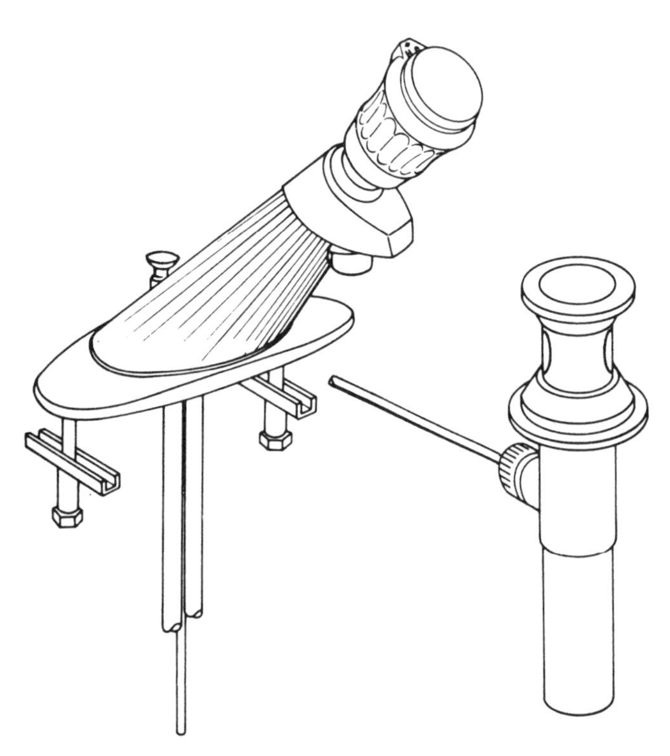

Replacing some types of strainers requires two people—one holds the unit securely in the sink while the other tightens it in place from below. With other types, all the work can be handled from underneath.

Thoroughly clean around the sink opening. Then apply a bead of plumber's putty all around before inserting the new strainer. After securing the strainer to the drainpipe, wipe away excess putty with a soft cloth.

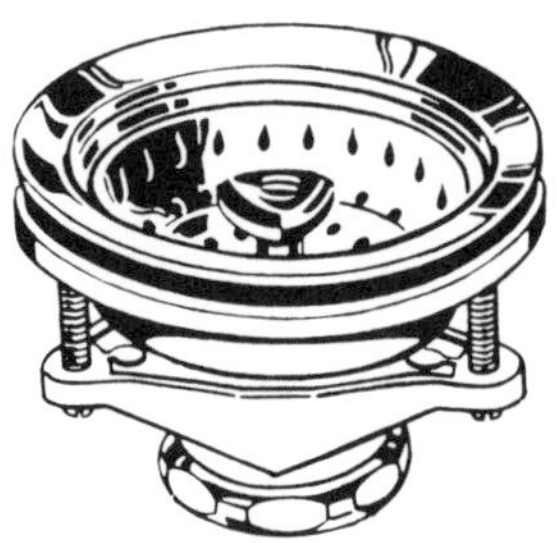

A strainer with a simplified mounting ring can be installed by one person.

Lavatory Faucets

Installation of lavatory faucets is similar to the installation of kitchen sink faucets, except that many lavatory units have pop-up drains controlled from the faucet deck.

Typically, such a unit has a plunger located behind the faucet spout. An adjustable lift rod is attached to this (beneath the lavatory), which moves a lever. The lever leads to a fitting in the drain that controls the up-and-down opening–closing motion of the drain plug.

To install such a unit, first remove the lift rod from the plunger. Wipe the sink clean and apply plumber's putty where necessary. Set the new faucet unit in place, with the plunger rod through the hole between the water supply holes. Tighten the slip nuts around the faucet shanks and connect the water supply lines, just as for a kitchen sink faucet.

Now remove the old pop-up drain and thoroughly clean around the drain opening. Apply a bead of plumber's putty around the opening. Follow the manufacturer's instructions for installing the drain. Fasten the lift rod to the plunger, adjusting it so that the plunger knob clears the faucet deck.

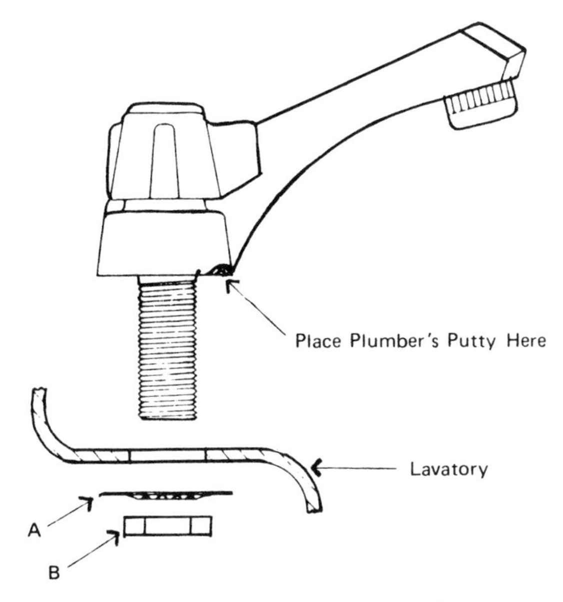

A typical lavatory faucet installation. Apply plumber's putty to the groove at the front of the deck. Insert the shanks of the faucet into the holes in lavatory. Attach washers (A) *and locknuts* (B) *to the shanks. Tighten with a basin wrench; then connect the water supply.*

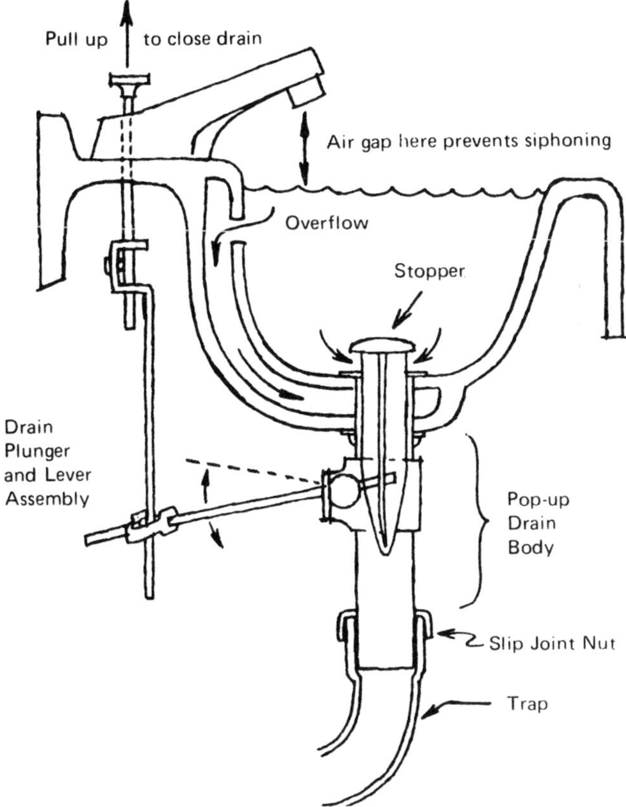

Operation of a typical lavatory faucet with pop-up drain.

Fixture Supply Tubes

The fixture supply tubes (usually flexible) that carry water from the supply lines to the faucets or other outlets are often replaced at the same time as faucets. They also may need to be replaced if they are damaged or develop leaks. In either case, the job is easy.

First, as always, shut off the water at the supply stop valve under the sink or other fixture or in the basement. Loosen the upper and lower connecting nuts that hold the tube in place. Place a pail beneath the tube to catch the water remaining in the line. Then remove the tube.

Tubing connections are made with a compression sleeve or nosepiece and a tightening nut. Place a sleeve over one end of the tubing and fit it into the upper connection; slide a nut on from the other end and finger-tighten

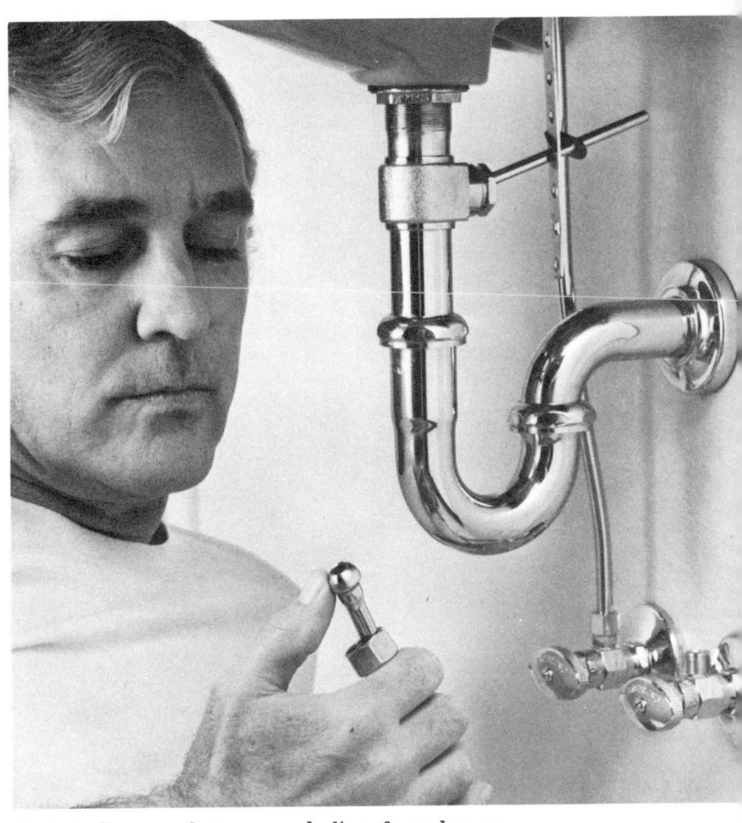

To install a new fixture supply line, first place a compression sleeve or nosepiece over the tubing and fit it into the fixture connection.

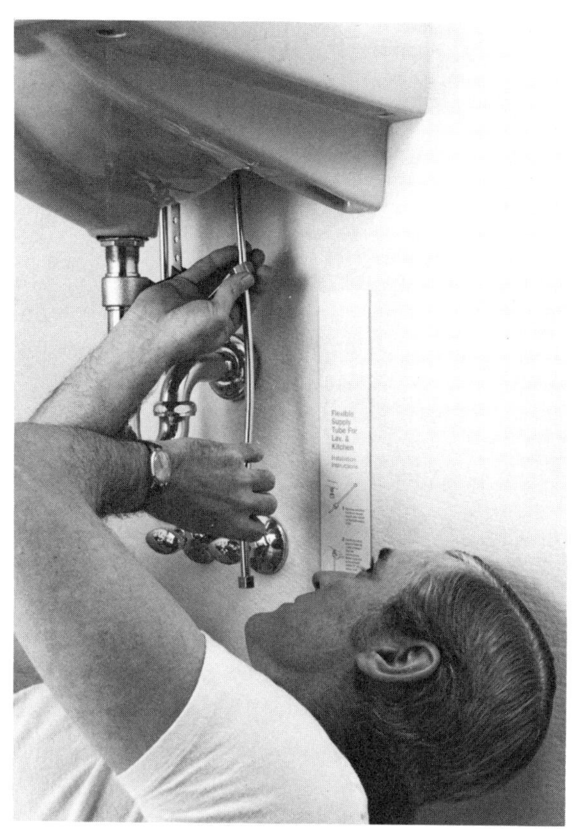

With nuts slipped onto the line, gently bend the tube to line up with the lower connection.

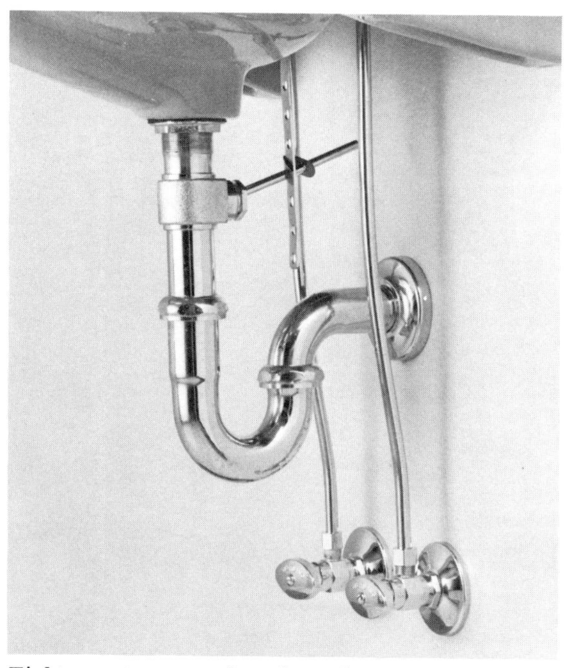

Tighten nuts to complete the replacement.

the connection. Gently bend the tube to line up with the lower connection; then slide the lower nut over the tube and slip it onto the sleeve. Bring down the nut and tighten it by hand. Finally, tighten both nuts with a wrench, but do not apply too much pressure—this will weaken the joint, and it may break the connection.

Supply Stops

The valves that control the flow of water to faucets and other outlets are not normally replaced when new faucets are installed. But they may become worn or damaged and have to be replaced. Whether these fittings are straight or at a 90° angle, the replacement procedure is the same.

Water must be shut off at a valve below the supply stop, usually in the basement or crawl space. Place a bucket below the valve to catch the water remaining in the line. Then unscrew the nut that holds the fixture supply tube to the supply stop.

If the supply line leading from the wall or floor into the supply stop is a threaded pipe, unscrew the old stop. Clean off the threads of the pipe and apply new pipe joint compound or tape. Then screw on the new stop and reattach the supply tube (or install a new tube).

If the supply line leading into the supply stop is soft copper with a compression fitting, replace the stop with a type having both a compression inlet and compression outlet. If the connection is a flared joint, check the flaring to make sure it is perfectly round before installing the new supply stop. If necessary, reflare the supply line (see Chapter 4, Flexible Copper Tubing).

Before replacing a supply stop with a soldered inlet joint, drain the water out of the line. You can do this by disassembling the old valve or by draining it at some previous point in the supply line. The new valve should be disassembled before soldering so that it is not damaged by the heat. Clean the end of the supply line with steel wool or an emery cloth; then apply flux to both the pipe and the stop fitting. Place the stop fitting on the pipe, give it a few turns to spread the flux

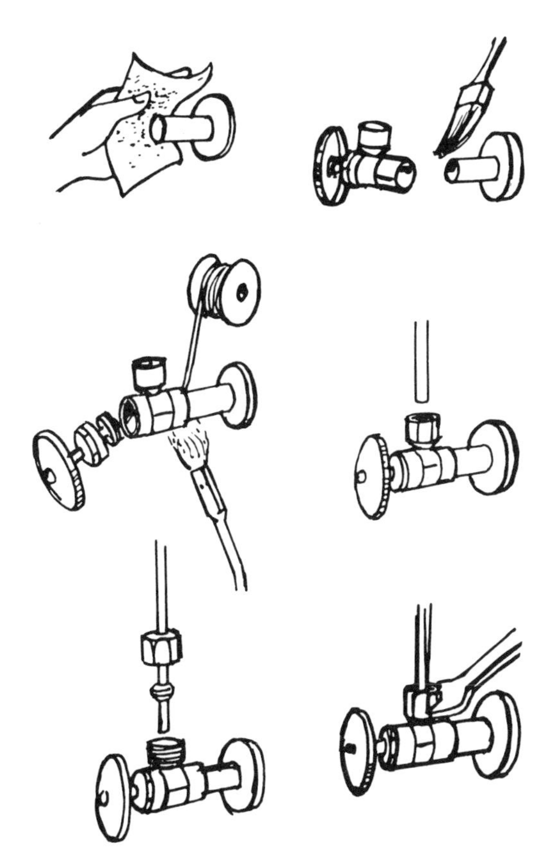

Installing a supply stop valve to a copper water line with a solder connection. Disassemble the valve before soldering to protect it from damage. A compression fitting attaches the valve to the fixture supply line.

evenly, and align the stop outlet with the faucet supply tube. Then heat the joint with a propane torch and solder (see Chapter 4, Rigid Copper Tubing). When the fitting has cooled, reassemble the valve.

Bathtub and Shower Faucets

If your bathroom features one of those big, old-fashioned, claw-footed bathtubs, despair not. You've got a head start on the new-fangled chic. More and more fixture manufacturers offer these freestanding tubs as part of their lines, along with the flush-with-the-wall types that predominated before the nostalgia wave swept the country.

Aside from questionable aesthetics or a yearning to return to the fancied good old

A three-valve tub and shower control (above) *and its typical installation* (below).

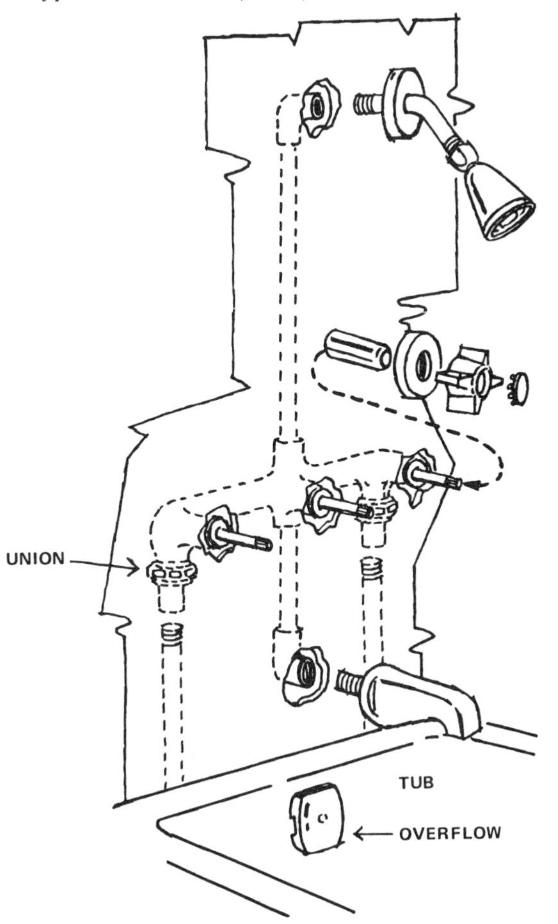

days, those venerable tubs did offer one big advantage: the plumbing was out there in the open for all to see—and repair, if necessary. Replacement of faulty faucets poses little problem. The project is essentially the same as installing new faucets in a sink or lavatory (only easier—you don't have to lie on your back while tightening connections above your head).

With more modern, or at least more conventional bathtubs, it's a different story. The piping is out of sight and out of reach when repairs or replacements are required—at least, so it seems. Actually, it is rarely necessary to tear out a bathroom wall to install new fittings. Most builders provide access doors or panels that enable you to get at the bathtub plumbing; in many localities, they are required to do so.

Before hacking away at the tile and plaster, look around for the access panel. Often, it is concealed in a closet adjacent to the bathroom, next to the tub wall. Once you have located and opened the panel, replac-

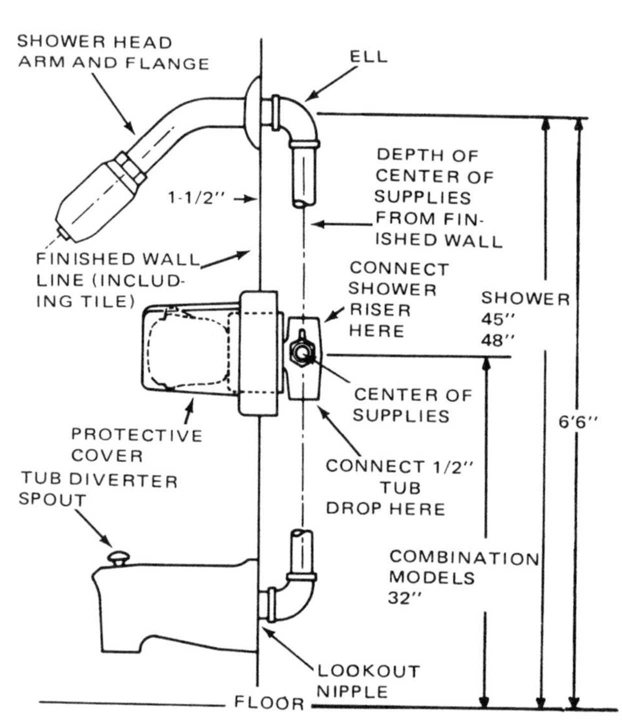

Left and above: *A single-handle tub and shower control with tub diverter spout.*

A shower head can usually be replaced quickly with a wrench, but be sure you protect the finish of the new head with tape where you will place the wrench.

ing bathtub fittings is similar to replacing other fittings previously described.

Bathtub Drains

The simplest type of bathtub drain control is the rubber plug (usually attached to the overflow drain with a chain) that fits into the

A pop-up tub drain is similar to a pop-up lavatory drain.

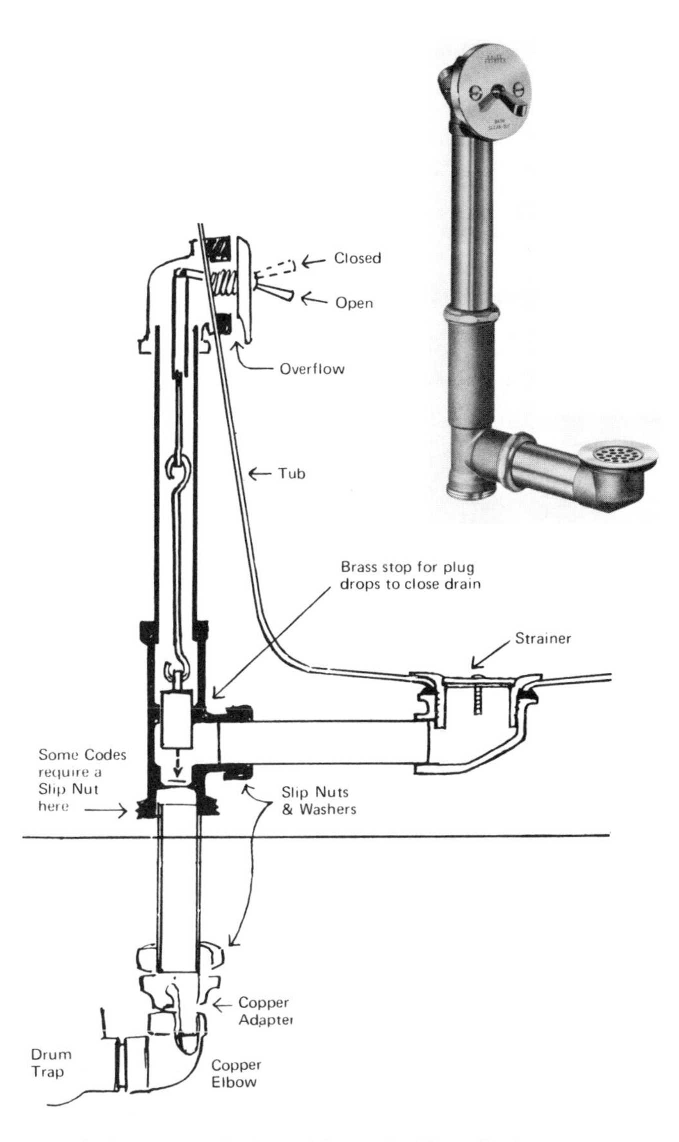

A trip-lever bath drain and its typical installation.

drain hole. Other than the deterioration of the plug—which you don't really need any plumber's skills to replace—the only parts that may possibly wear and have to be replaced are the drain line and the overflow tube. Both are readily accessible from the rear access panel (see above).

Slightly more complex are pop-up and trip-lever mechanisms controlling the drain opening. Because of the moving parts, these are susceptible to wear and may have to be replaced. Again, the access panel makes this a relatively simple task. Adjustable lift rods are provided to fit any size tub.

Sink Traps

Sometimes traps under sinks and in some first-floor tub installations where the trap is exposed in the basement must be replaced because the fitting has corroded or a bad leak has developed at the joints.

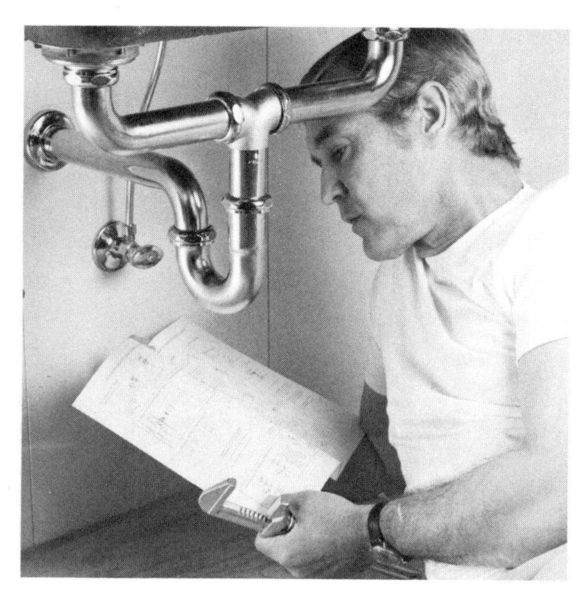

Before replacing a trap bend, first remove all nuts and seal washers with the old bend.

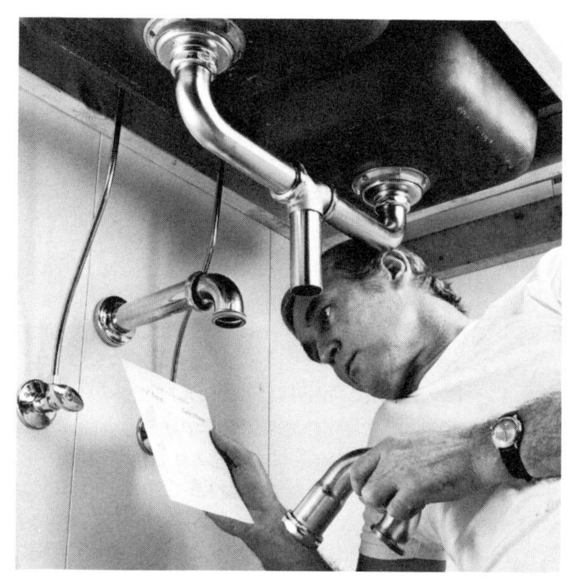

Place the nuts from the new bend over the existing pipes; then fit the new bend to the sink drain and wall outlet.

Place a bucket below the old fitting to catch the water remaining in the trap bend. Remove the nuts and the seal washers and pull off the trap.

Place the nuts and the washers from the replacement trap over the sink tailpiece and the outflow pipe (place the nuts on first, then the washers). Slide the long end of the replacement trap over the sink tailpiece and fit the other end onto the outflow pipe. Slide the washers and the nuts down to the replacement trap and tighten them with an adjustable open-end or monkey wrench.

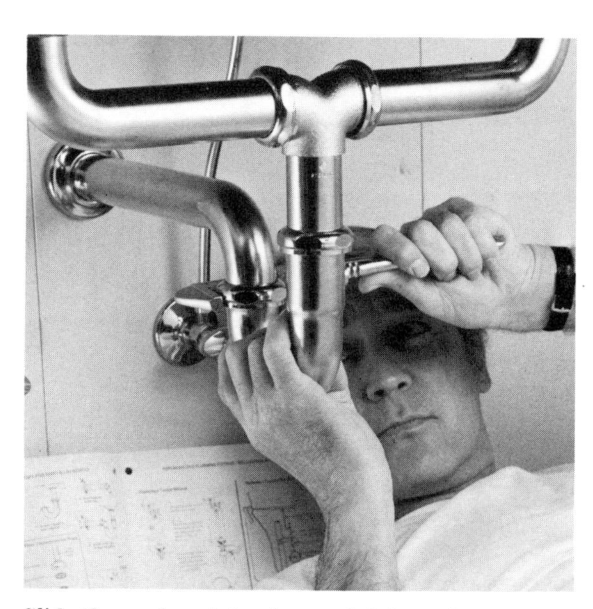

Slide the washers into place and tighten the nuts with an adjustable wrench.

Toilets— The Mysterious Mechanism

It has been a hundred years since Thomas Crapper invented the flush toilet, but it still remains a mysterious mechanism to many people. Other than jiggling the handle and, possibly, bending the float rod when the device refuses to stop running, they stand in awe at the workings of its hidden innards. And when things go wrong—as things always will when they involve moving parts—it is the plumber who usually benefits from this toilet mystique.

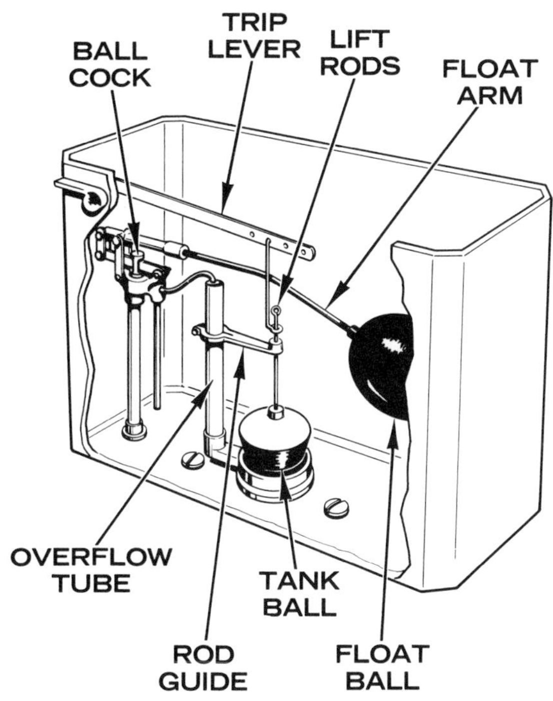

Exposed—the secrets of the toilet tank! And it turns out not to be so formidable a mechanism after all. The overflow tube, guide, lift rods, and tank ball are parts of the flush valve. The float arm and float ball go with the ballcock.

But wondrous as it may seem, the flush mechanism of a toilet is quite simple. Basically, it consists of two assemblies: a flush valve and a ballcock. Individual parts of each assembly may wear out and have to be replaced (for example, the tank ball and float), but eventually the entire assembly or assemblies will probably require replacement. This is not a major job.

First empty the water from the toilet tank. This is done by shutting off the water at the supply stop (or some other valve in the line) and then flushing the toilet (holding the handle down until most of the water drains out of the tank).

Replacing a flush valve usually involves removing the old flush valve along with the lift wires, the washer, and the locknut that holds the valve seat in place. The discharge tube of the new valve assembly is inserted through the tank bottom. After the overflow tube has been positioned, tighten the locknut. This locks the discharge tube in place. The guide arm fastened to the overflow tube is centered over the valve seat and is tightened. Install the lift wires through the guide arm and the trip wire. Then screw the flush ball onto the lower lift wire, aligning it so that it will drop into the exact center of the flush valve seat.

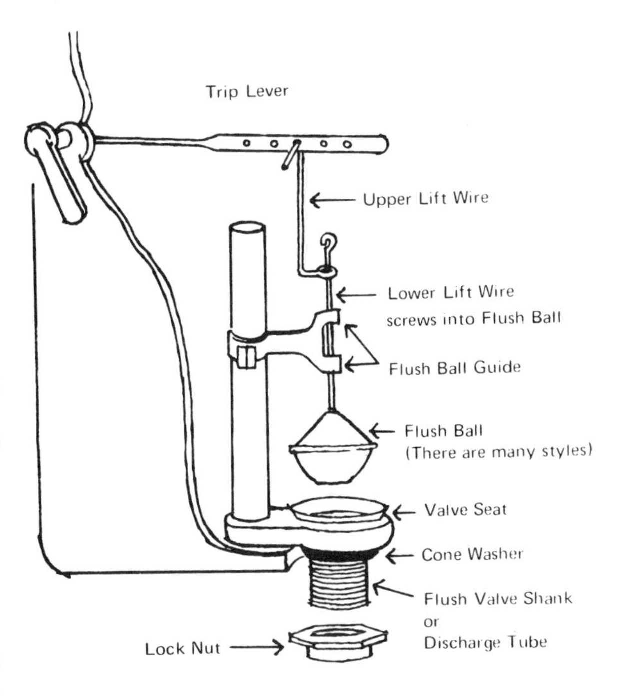

To replace a flush valve, remove the old valve and insert the new discharge tube through the tank bottom. Situate the unit so that the overflow tube is properly located; then tighten the locknut under the tank. Position the guide arm and install the lift wires and tank ball.

Replacing a ballcock is no more difficult than replacing the flush valve. In fact, replacing the whole is considerably easier than replacing any of its parts, except possibly the float and float rod (see Chapter 9). First empty the tank (see above). Then place a pail under the supply connection to catch the water that remains in the tank. Unscrew the fixture supply tube and the locknut from the ballcock shank and remove the old assembly. Remove the locknut from the new ballcock and place the shank through the opening in

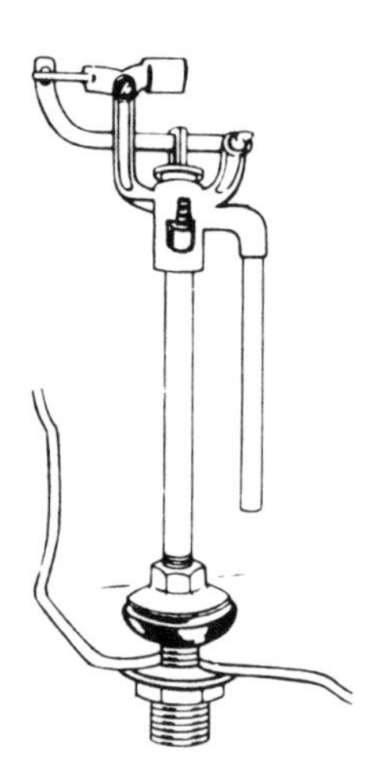

To replace a typical ballcock, place the shank of the new ballcock through the opening in the tank with the large rubber washer inside the tank.

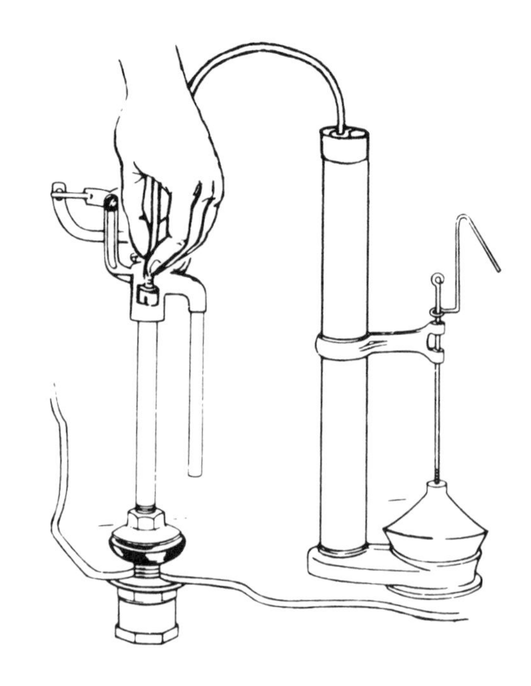

Slip one end of the refill tube over the lug on the body of the ballcock and place the other in the overflow tube.

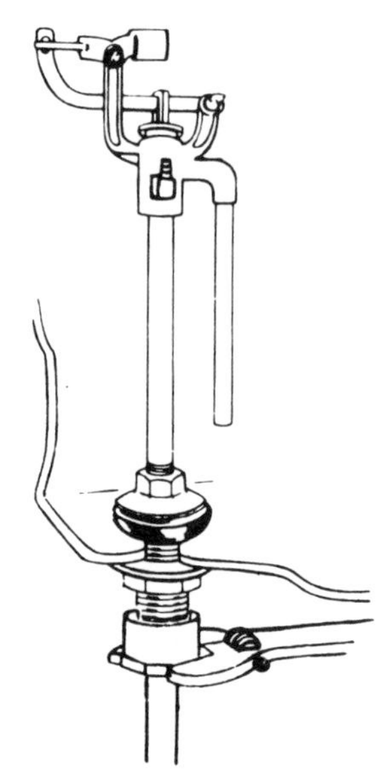

Connect the water supply line to the shank.

the tank, with the large rubber cone washer inside the tank. Tighten the locknut over the shank; then fasten the supply line. Connect the refill tube (usually plastic) to the ballcock and place the other end into the overflow tube of the flush valve assembly. Screw the float rod into the ballcock assembly and fasten the float to the end of the rod. Turn on the water, allowing the tank to fill. It may be necessary to bend the float rod slightly—and very gently—to permit the tank to fill properly and to keep the rod clear of the overflow

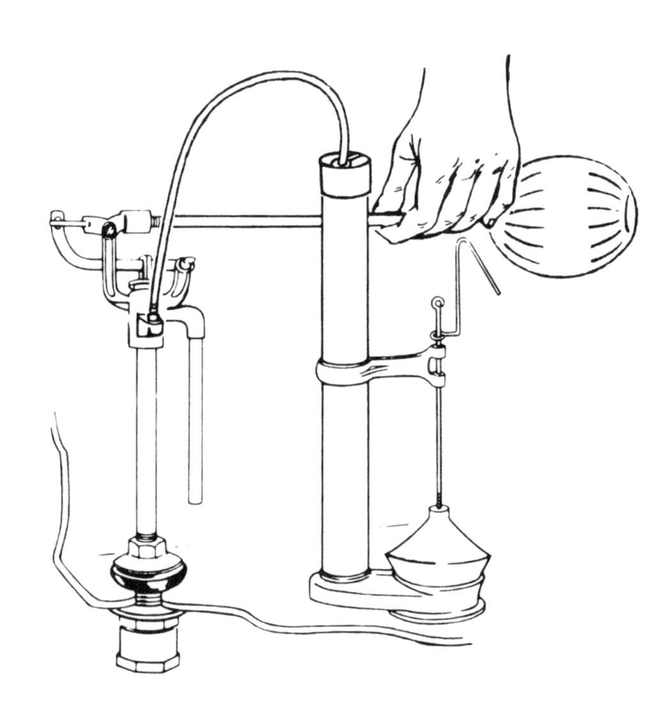

Assemble the float rod and float ball with the ballcock. Bend the rod slightly, if necessary, to clear other tank parts and to make the tank fill properly.

tube and other parts of the mechanism.

A fairly new type of ballcock on the plumbing market eliminates the need for the float and float rod, thus eliminating the source of a lot of toilet troubles. Since this device works on a different principle than conventional ballcocks (fluid energy), the tank is filled faster and with less noise. It is designed to come on full flow when only one gallon of water is lost, alerting you to the fact that something is wrong. The trouble (usually a defective tank ball) is quickly spotted and fixed. Without this feature, the water might leak out slowly for many weeks before it became pronounced enough to be noticed. With the price of everything (even water) still rising, it helps to have this early warning system. The installation of the floatless ballcock is similar to the installation of the conventional unit (minus the float rod and float, of course).

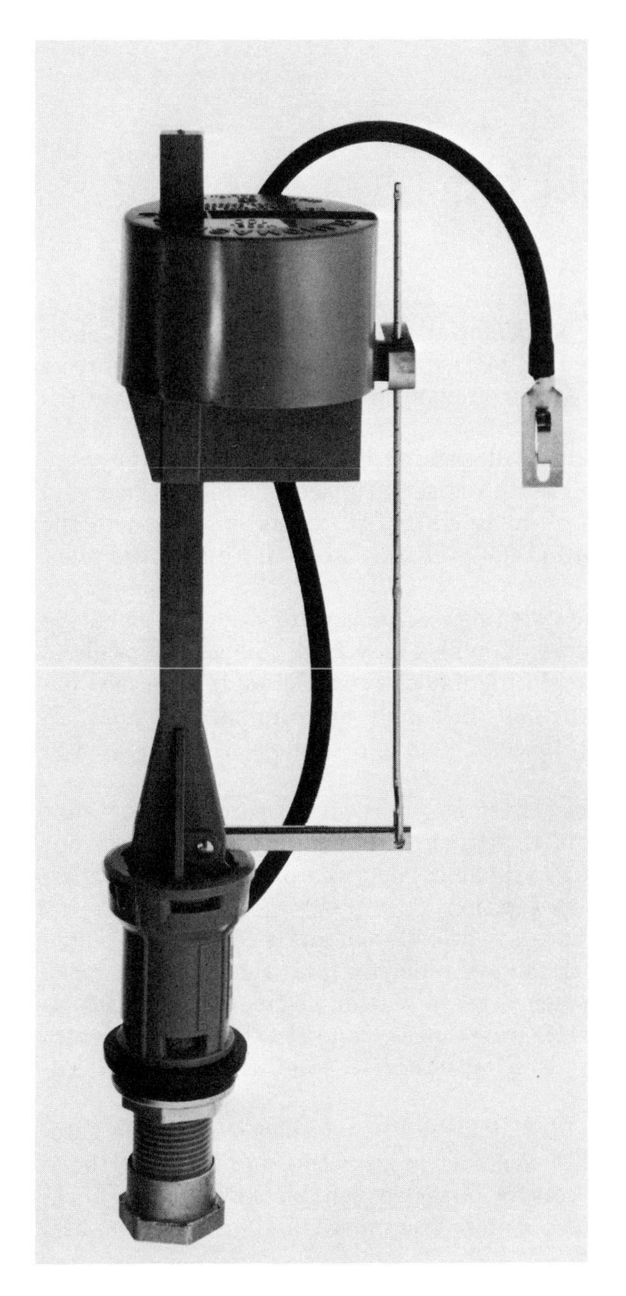

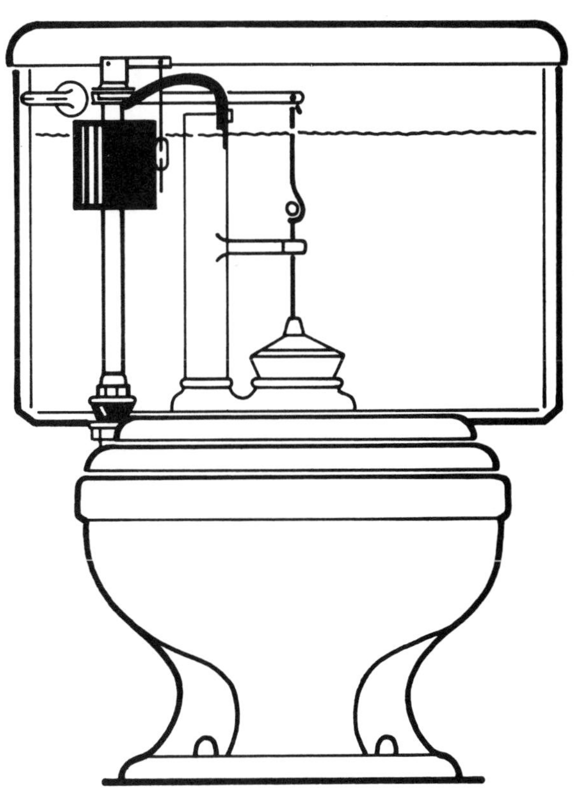

Left and above: *A new type of ballcock eliminates the float rod and float ball, simplifying installation and operation.*

7
Selecting and Installing Fixtures

Good plumbing fixtures last a long time, but eventually they will have to be replaced. The fixtures may be damaged, or they may just plain show their age. Perhaps you may need to replace fixtures as part of a bathroom remodeling project. The needs of your growing family even may dictate the addition of a new bathroom.

In a new installation, where the wall framing is still exposed, supply and drain lines are roughed-in first. The tub is set in place before the floor and walls are finished. Backing for wall-hung sinks and toilets is secured to the wall studs; then the wallcovering and floor tile are installed. Finally, the toilet and lavatory are installed.

If you are replacing old fixtures with new ones, you may have to open the walls to install the tub and lavatory. You also may have to open the walls to install a wall-hung toilet if it is replacing a floor type. Before buying new fixtures, check the roughing-in dimensions of supply and drainage openings. If they are the same as or close to those of your old fixtures, it will save you a lot of work.

Don't skimp when you buy new fixtures—it is an investment you won't make very often, so you can afford to let factors other than price affect your choice. A multitude of designs are available, each one offering unique characteristics that warrant your consideration.

That doesn't mean that you have to go for broke and select a gold-plated bathtub. Most fixture manufacturers have economy lines that are well made and can be counted on to give many years of trouble-free service. But as the fixtures move up toward the top price ranges, you get extra comfort, extra convenience, and extra elegance. You must decide whether the added features are worth the added cost.

After your new fixtures have been delivered, treat them gently. Be careful not to chip or scratch finished surfaces when uncrating and handling them. Rest the fixtures on pads if necessary. Use wooden blocks to brace them temporarily in position. If, for some reason, you must stand on a fixture, first take off your shoes.

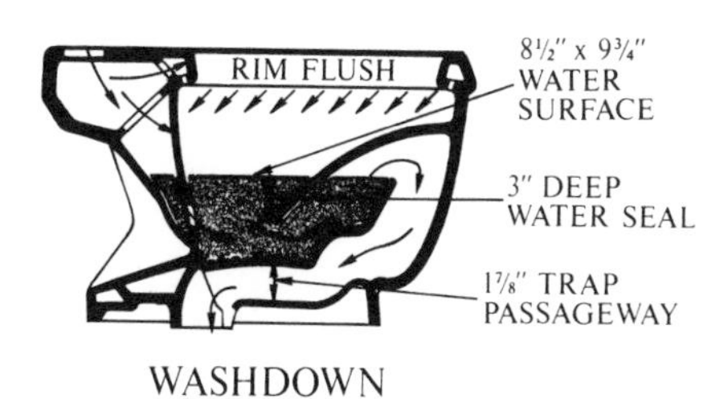

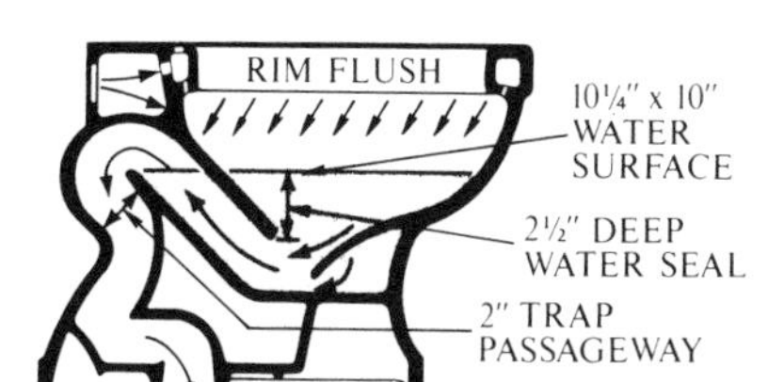

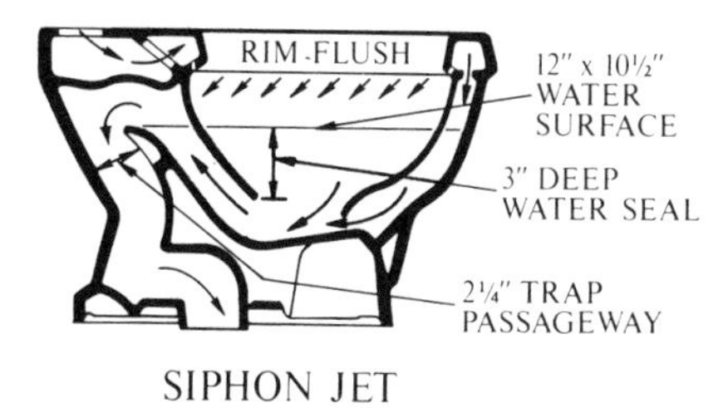

Three types of toilet action: a larger water surface area means less staining and fouling; a larger trap passageway reduces the chance of clogging.

Toilets

Most residential toilets consist of a bowl and a tank. The two may be separate or in a one-piece unit. The function of the tank is to store enough water for proper flushing action. The mechanisms inside the tank (the ballcock and flush valve) were discussed in Chapter 6.

All toilets remove waste, but some do it better than others. A good toilet should operate quietly and provide a large water surface area to prevent contamination and fouling. The four basic kinds of residential toilets can be distinguished by their different flushing actions. Make sure you know which one you are buying.

Washdown toilets are the least expensive, the least efficient, and the noisiest. They are flushed by a simple wash-out action and may clog more easily than other types. Much of the bowl area is not covered by water and is subject to fouling, staining, and contamination.

Reverse trap toilets are flushed by creating a siphon action in the trapway, assisted by a water jet located at the inlet to the trapway. This siphon pulls the waste from the bowl. It is moderately noisy but efficient, and it is the least expensive of the siphon-action toilets. It is less likely to clog than the washdown type, and since more of the interior bowl surface is covered with water, it is subject to less fouling.

Siphon jet toilets are an improved version of the reverse trap bowl. They have a larger

Special fixtures suit special purposes. Here, a corner lavatory (reflected many times in the wall mirrors) and a corner toilet (with storage cabinet above) turn a tiny 4-foot-square space into a neat powder room.

This one-piece low-profile toilet operates on a siphon vortex principle and has an almost silent flushing action.

A round-bowl toilet can be used where space is at a premium, but an elongated bowl is preferred for comfort and cleanliness.

A wall-hung toilet (here with an adjacent bidet) makes floor cleaning easy.

water surface; most of the bowl interior is covered with water. The trapway is larger, making the flushing action quieter and less subject to clogging than that of the regular reverse trap toilet.

Siphon vortex toilets are low-profile, one-piece toilets with a near-silent flushing action and almost no dry surfaces on the interior of the bowl. Efficient and attractive, this is the most expensive of the toilet types.

Most toilets are available with either round or elongated bowl rims. You probably will prefer an elongated bowl, which most people find more comfortable, more attractive, and (since it provides a larger interior water surface), easier to keep clean.

Off-the-floor toilets are attractive and efficient, and they give the added benefit of easy floor cleaning. But they are difficult to install in existing bathrooms since they require the placement of special metal carriers inside the wall to support their weight. In new construction, this is no problem.

There also are special designs to fit particular situations. For example, a toilet with a triangular tank makes possible a corner installation, taking advantage of space that might otherwise be wasted.

Toilet Installation

To remove an old toilet, first shut off the water, either at the supply stop or at some other valve in the line. Then flush the toilet, holding the handle down until most of the water drains out of the tank. Sop up the water remaining in the tank with a sponge or rags. Disconnect the water supply tube at the bottom of the tank.

If you are working on a two-piece toilet, remove the bolts that hold the tank to the bowl (they are usually located inside the tank). Then lift off the tank. If the tank is mounted on a bracket fastened to the wall, disconnect the pipe between the tank and bowl and lift the tank off the bracket.

Pry off the covers from the bolts that hold the bowl to the floor and remove the nuts. Twist the toilet bowl slightly to free it; then lift it straight up. Be careful not to tilt it backward in case some water still remains in the bowl.

Remove the old bolts from the floor and the floor flange. Use a putty knife to scrape away any old wax or other material from the flange. After it has been wiped clean, insert two new bolts in the slots of the flange.

If you are installing a new toilet rather than replacing one, install a toilet flange fitting into the waste pipe in the floor. The flange surrounds the waste pipe and screws to the floor; it includes slots for the mounting bolts. Insert two bolts in the flange. Some toilets require only these bolts; most require two more toward the front of the bowl. Mark the location for the front bolts, using a paper template or the bowl itself. Drive hanger bolts (bolts with wood threads at one end and machine threads to take nuts at the other) into these spots, making sure you keep them perfectly vertical.

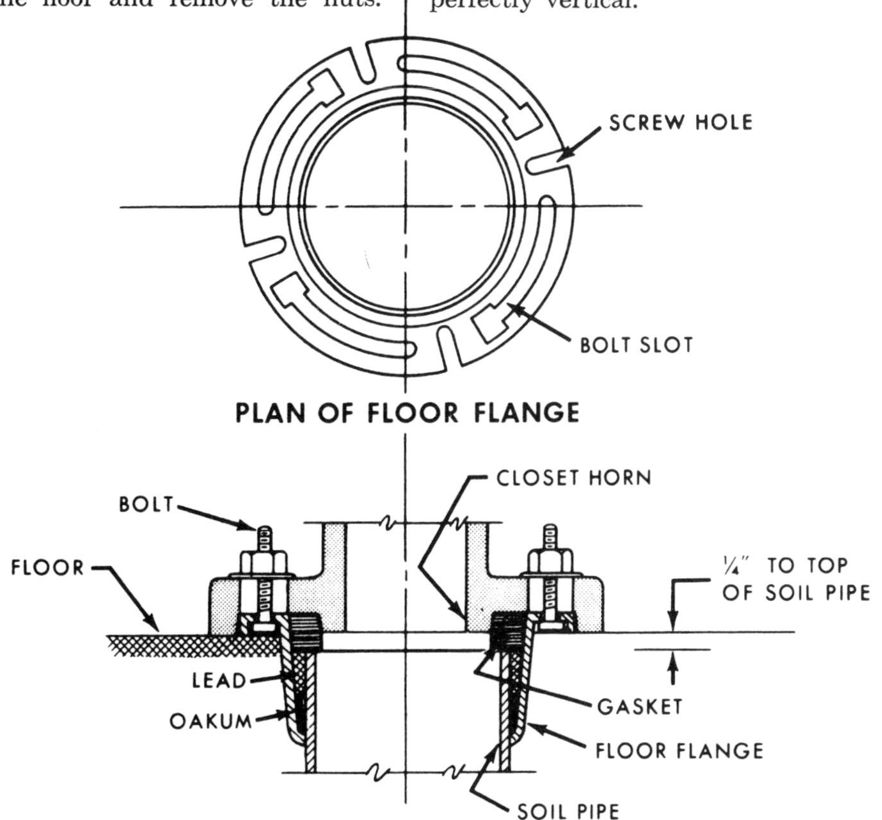

Installing a toilet. Step 1: Toilet flange is screwed to the floor; then it it is fitted with bolts to anchor the toilet bowl.

Step 2: A putty ring is applied around the rim of the toilet base, and a rubber or wax gasket is fitted to the discharge opening.

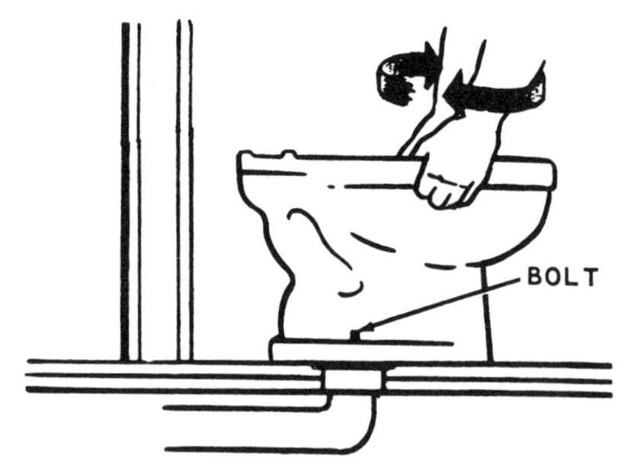

Step 3: The toilet is set in place over the bolts and the floor drain; then it is turned slightly to make a firm seat.

Step 4: The toilet tank is fitted on top of the bowl and is bolted into place.

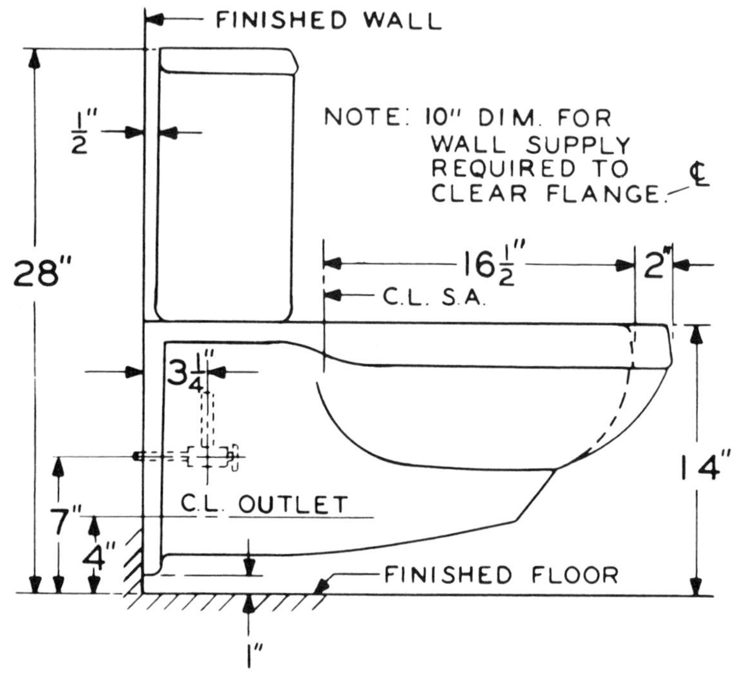

Wall-hung toilets discharge through the rear. Special carriers must be attached to the studs within the wall to support the toilet's weight.

Place padding or newspapers on the floor and turn the bowl upside down on them. Lay down a bead of plumber's putty completely around the inner edge of the base rim. Place a new wax or rubber gasket around the discharge opening.

Lift up the bowl (with a helper, if necessary) and set it down gently with the discharge opening over the floor flange and with the flange and floor bolts through the holes in the base. Let it down into final position as straight as possible—do not disturb the putty ring and gasket. Press down on the top center of the bowl and twist it very slightly to settle it into the putty. Check to make sure that the bowl is perfectly level. If necessary, use wood wedges to level it, but be sure that the wedging does not lift the bowl up enough to leave air gaps in the putty ring or the seal around the discharge opening. When the bowl is squarely seated, place the washers and nuts over the bolts. Tighten them down to a snug fit, but do not force-tighten the nuts. Place a small amount of putty inside the bolt covers (which are usually made of ceramic or plastic) and set them over the exposed bolt ends.

If the toilet is a two-piece unit with a separate tank, the tank is usually bolted to the bowl. Place a rubber gasket over the water connection; then set the tank in place and tighten the nuts. Complete the installation by connecting a new fixture supply line to the tank (see Chapter 6).

Bathtubs

Bathtubs come in a variety of sizes and shapes. If you are replacing an old tub, you are restricted to installing a new one that is similar—unless you care to have some extensive carpentry done (for example, a few walls may need to be knocked down to change the shape of the room). For a new bathroom, you have a wide choice.

The most frequently used type of bathtub is rectangular. It is usually recessed into a niche, which is then tiled or finished with some other water-shedding material that can withstand showering. Rectangular corner bathtubs are made for installations where the back and only one end of the tub will be against walls. This model is more expensive.

Of the many other unique bathtub shapes that are manufactured, the square tub is most popular. Premium-priced, a full-size square bathtub requires more floor space than the rectangular models. Small square tubs work well in areas where space is a problem and are fine for children's baths, but they are not large enough for adult bathing.

Tub sizes vary widely. The length of a standard rectangular tub ranges from 4 to 6 feet. Heights range from 12 to 16 inches from the floor to the top of the rim. The depth of water in the tub is governed by the distance from the tub bottom to the overflow outlet. This dimension is not always given in manufacturer's literature, and you may have to obtain it by measuring a sample in a dealer's showroom.

Most bathtubs are manufactured from one of three materials: cast iron with a porcelain enamel surface, formed steel with a porcelain enamel surface, or a molded gel-coated glass fiber reinforced with polyester resin. Units made of the latter material are generally known as *fiberglass bathtubs.*

Cast-iron bathtubs were first manufactured in 1870, but the enameling process dates back to ancient Egypt and Assyria. The combination of thick, glossy porcelain and the heavy rigid body makes these tubs less susceptible to damage than tubs of other materials. They come in lengths of 4, 4½, 5, 5½, and 6 feet, in widths from 30 to 48 inches, and in depths from 12 to 16 inches. They are heavy, weighing as much as 500 pounds.

Some cast-iron tubs have a short front apron so that they can be recessed into the floor for a sunken tub installation. Sunken tubs pose something of a safety hazard, however, and rails or grab bars should be provided.

Formed steel bathtubs with a porcelain enamel finish are lightweight (about 100 pounds) and less expensive than the cast-iron tubs. They are ideally suited for upper-story installations or for remodeling because of the ease with which they can be moved into place and because they normally do not require extra floor reinforcement (as is sometimes necessary for a heavier tub).

Some steel tubs are made in two pieces, with the apron formed separately and welded into place before enameling. Since the welded seam may be visible, one-piece tubs are preferred.

Most manufacturers of steel tubs offer a sound-deadening undercoating, sometimes as an extra-cost option. The reduction of shower noise justifies this slight additional cost. Steel tubs are usually available in 4½- and 5-foot

Traditional cast-iron tub with a porcelain enamel finish is recessed into a wall niche.

lengths, in 30- to 36-inch widths, and in depths of 15 and 15½ inches.

Fiberglass bathtubs have come into widespread use only in the last decade. The surface is a smooth and attractive gel coat, but it is not as durable as porcelain. While it is extremely easy to clean, abrasive cleaners should not be used since they will harm the finish.

One-piece fiberglass tub–showers combine bathtubs and the surrounding walls (up to about 73 inches above the floor). This solves the problem of maintaining a neat, watertight joint where the tub and the surrounding walls come together. There are many variations in size and design, but those most commonly used in homes are the 5-foot-long rectangular models.

Fiberglass bathtubs are lightweight, making them ideal for upstairs bathrooms.

Some cast-iron bathtubs have a short front apron for sunken-tub installation. Although not shown in this illustration, grab bars should be provided for safety.

One-piece fiberglass tub–shower units eliminate the joint between the tub and surrounding walls, making installation easy.

Bathtub Installation

If you are planning to bury your new bathtub in the middle of the floor to provide an exotic setting for sunken bathing, you'd better call in an architect or an engineer for advice before you begin. But more typical tub installations are relatively simple (except for wrestling a 500-pound cast-iron monster into position).

When installing a new tub in an existing bathroom, the plaster or other wall covering must be removed around the tub area.

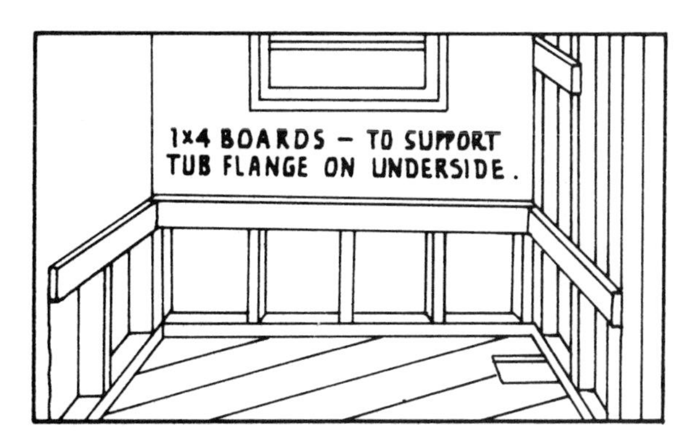

For steel or fiberglass tubs, 1 × 4 flange-support boards are nailed to the surrounding wall studs.

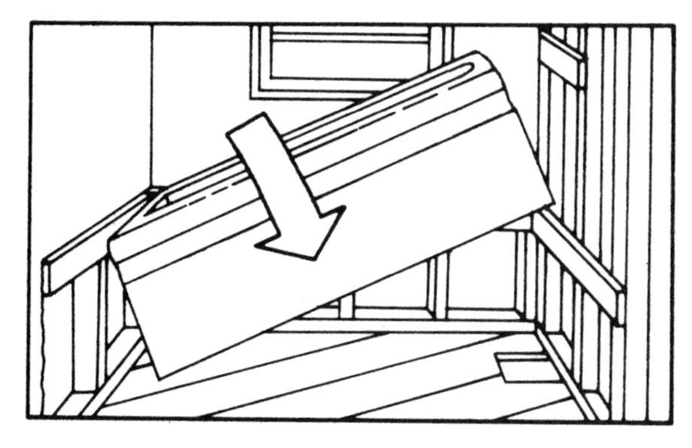

The new tub is fitted in place and the walls are re-covered before final plumbing.

When the new tub will be located in a niche —completely filling the space—the wall covering should be removed completely down to the studs from one end wall (the plumbing wall). This allows some leeway when lowering the tub into place. Remove the wall covering from the other end wall and the side wall to the height of the tub. (If the niche is longer than the new tub, open only one end wall and the side wall; then build out from the other end to fill the extra space.)

If you are installing a formed steel or fiberglass tub, nail 1 × 4 boards to the faces of the studs. The tops of the boards must be perfectly level and exactly at the height required to support the tub flanges (cast-iron tubs normally do not have such flanges). Lower the tub into position with the flanges resting on the 1 × 4s; anchor the tub by inserting screws through the flange holes into the boards.

If the shower plumbing is already in place (as it should be from the previous installation or from the roughing-in operation), the walls now can be refinished to make a neat fit around the new tub. Supply and drain connections should be accessible through a rear access panel and are connected as described in Chapter 6.

To install a one-piece fiberglass tub–shower as a replacement for a bathtub in an existing bathroom, remove the wall covering on all three sides of the niche from floor to ceiling. (Or, if the new unit does not fill the space, remove the wall covering on two walls and build a stud wall to support the third side of the tub–shower.)

Set the unit in place, using wedges to make it level if necessary. Nail the flanges to the wall studs. Then apply furring strips to the studs above the unit's walls, running them up to the ceiling. Apply water-resistant sealer to the face of the nailing flange, then cover the flange by nailing wallboard to the furring strips. Paint the wallboard to match the rest of the room.

Installation of fiberglass tub–shower is a one-man job. Here it is being installed in a new construction; but it can be put into an existing room as well.

Prefabricated shower units make it possible to install a shower in a small space or in a room where there was no shower before.

Larger shower units have comforts such as molded-in seats.

Shower Stalls

One-piece fiberglass shower stalls are just the thing for the do-it-yourself remodeler of an older bathroom which never enjoyed a shower, or for those wishing to install a small new bathroom. They come in many sizes, but they all take up much less room than a bathtub. For comfortable adult showering, a 30-×-36-inch stall is the minimum adequate size. Bigger is better, if you can afford the space. Some larger sizes even come with seats molded in.

Installing a shower stall is similar to installing a one-piece tub–shower combination, as described above. If the new unit is to be placed in a corner, a stud wall is built along the third side; if it is to have only one side against an existing wall, stud walls must be constructed along the remaining two sides of the shower stall.

Lavatories

Lavatories come in so many different styles, shapes, and sizes that you are certain to find one that is perfectly suited to your family's needs and your aesthetic sensibilities, subject only to the amount of space available.

Most lavatories are made of cast iron, vitreous china, or formed steel. The techniques of casting (iron) and molding (china) allow great license in shape and design. The forming process used in the manufacture of steel lavatories places more limitations on the final product. However, in many installations, it is difficult to notice any difference among the three materials. A relative newcomer is the plastic ("synthetic marble") integral lavatory and countertop. The clean, seamless lines make this an attractive choice for the modern bathroom.

Among the more popular lavatory styles are the flush-mount, self-rimming, under-the-counter, and wall-hung types.

The flush-mount lavatory requires a metal frame to hold it in place. It is relatively inexpensive, but it is difficult to clean between the frame and the sink and between the frame and the counter in which it is mounted.

The self-rimming lavatory projects above the countertop and requires no framing ring. The sink rim rests in a bed of sealant on the counter. This type of sink offers easy cleaning.

The under-the-counter lavatory mounts beneath an opening cut in the countertop. The fittings are mounted through the countertop. This type of sink is difficult to clean near the seam where it joins the underside of the counter (where dirt is likely to collect).

The wall-hung lavatory is usually rectangular in shape, although it may have a rounded front or be triangular for fitting in corners. This type of sink is a space-saver, since it usually does not have a surrounding counter. However, under-counter storage space is sacrificed, and the plumbing is exposed.

Flush-mount lavatories are held in a bathroom counter by means of a metal frame. Round and D-shaped versions are pictured here.

Under-the-counter lavatories have fittings installed through countertop.

Above and below: *Self-rimming lavatories project above the countertop. They also come in many shapes, including round and the rectangular unit shown here, with large, gradually sloping basin, off-center fittings, and a built-in soap or lotion dispenser.*

This one-piece lavatory–countertop is mounted on a cabinet.

A wall-hung lavatory is suspended from a metal hanger or hangers fastened to a support that is nailed to wall studs.

Lavatory Installation

Installation techniques for lavatories are as varied as the types of units; for counter-mounted models, much depends on the type of counter, as well as on the type of lavatory.

A typical wall-hung lavatory requires behind-the-wall backing. To install such a unit in an existing room, you must cut through the plaster or other wall covering to get at the studs (unless you are replacing a sink that had

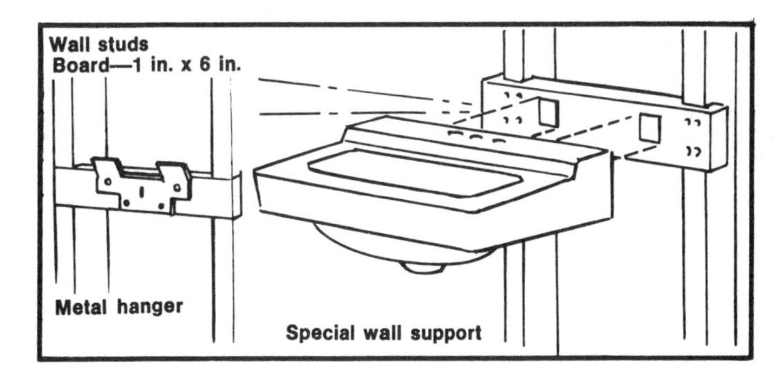

a similar backing). Firmly fasten a 1 × 8 board into ¾-inch-deep notches cut in the studs at lavatory height. The wall then can be refinished. Brackets then are anchored to the backer board and the lavatory is hung on the brackets against the wall. If there are legs at the front of the lavatory, they are adjusted easily to rest squarely on the floor.

For methods of installing lavatory fittings and traps, see Chapter 6.

Some wall-hung units include front legs for additional support.

Kitchen Sinks

Kitchen sinks also come in a variety of designs (although not in as many designs as bathroom lavatories), including single, double, triple (the third well is usually a small, shallow bowl flanked by two large sinks), deep-well, deep-and-shallow combination, and large-and-small combination styles. They are usually of porcelain-enameled cast iron or of formed or stainless steel. And they are almost invariably mounted in cabinets; the type of cabinet determines the method of mounting. Probably the most common is the flush-mount, which employs a metal frame to hold the sink in place, as described above. Stainless-steel sinks are often of the self-rimming variety.

Plumbing connections work on the same principles as those described for bathroom sinks in Chapter 6.

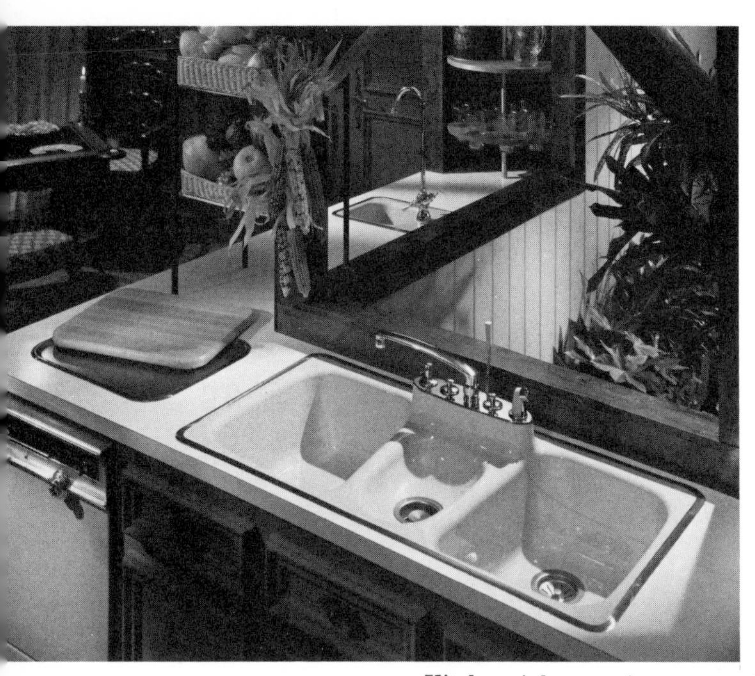

Kitchen sinks come in many configurations. Shown are a double unit with integral drainboard (top), *a triple-basin unit* (left), *and a sink with deep and shallow wells* (right). *All three are flush-mounted.*

Stainless steel is a popular choice for kitchen sinks. Most stainless steel units are self-rimming.

Laundry Tubs

The stationary laundry tub is a plumbing fixture generally found in the basement or a utility room near the kitchen. Stationary tubs also are available in a wide range of sizes and materials. Perhaps the best choice to stand up to the harsh bleaches and chemicals often used on washday is an enamel-finished cast-iron tub. It is easily cleaned, and the porcelain finish holds up well. They come in single- and double-compartment styles, and some even can be installed in a countertop. The method of installing a laundry tub is similar to that for installing other types of sinks. However, when the stationary tub is located in the basement, you can use ordinary pipe and fittings rather than the more expensive chrome-plated ones.

8
Planning a Modern Bathroom

Many plumbing fixtures are not the central features of the rooms they are in. For example, a sink and dishwasher, while very important to the kitchen, are not the *only* important functional items in that room. The range, the refrigerator, and adequate working space are at least equally vital to the storage, preparation, and serving of food.

But the bathroom is designed around basic plumbing fixtures: the toilet, the lavatory, and the bathtub and/or shower. Without these fixtures, the bathroom is not functional. Thus, a knowledge of plumbing and how it works is essential when you plan an updated or brand-new bathroom.

The bathroom used to be designed strictly for function. Today, the antiseptic white has given way to decorator colors as well as floor, wall, and ceiling materials that would be at home in any room. Still, the basic function of the room remains, and it is with this function in mind that plans must begin.

Whether you are adding a new bathroom or remodeling an existing one, you must anticipate future family needs, as well as consider present requirements. It may be wise to put in extra facilities now (or at least rough them in, depending on your budget) rather than adding them—at far greater expense—at some future time.

The bathroom, if there is only one in the house, should open into a hallway that makes it easily accessible to all rooms. Its entrance should not be visible from the front door or the living room. Ideally, it should be placed so that children can reach it from the back door without going through too many rooms.

If there are two bathrooms or an extra half-bath, an economical plumbing layout is to have them back-to-back or, in a two-story plan, to have one directly over the other. This helps minimize the amount of piping needed for the supply and drainage systems.

The Importance of Layout

A good arrangement within the bathroom is also important.

Compartmented or divided bathrooms can provide privacy for two or more people at the same time, and they require less space than two rooms. The

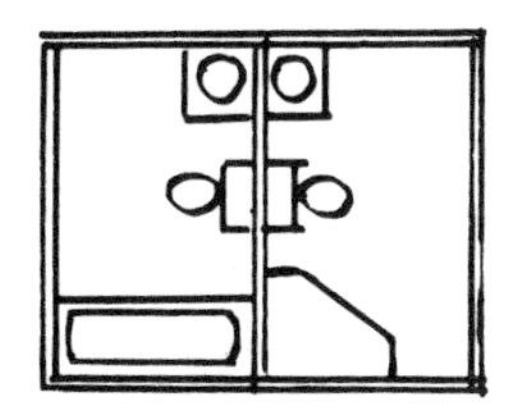

Installing fixtures back-to-back is most economical, but this one-time saving may be overruled by other considerations in your bathroom planning.

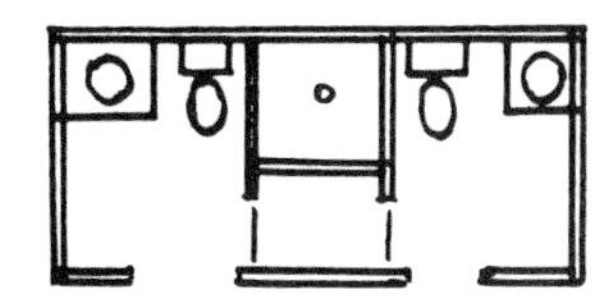

Here, one shower stall serves two lavatory–toilet areas.

Above and right: *Renovating an old bathroom will add comfort, convenience, elegance, and value to your home. Here, 50-year-old fixtures were removed. A new cast-iron tub was sunk in the floor, and a partial partition was built to screen the toilet. Two lavatories replace the old washstand. The ceramic tile floor, new draperies, the mirror and the light fixture complete the modernization.*

sink may be placed in one compartment, with the tub and toilet in the other. Or the toilet may have a compartment all its own.

When there is not enough space for two bathtubs, it may be possible to use one tub to serve two bathrooms. Or the bathroom may be divided into a series of separate rooms—one for the toilet, one for bathing facilities, and one for dressing and grooming.

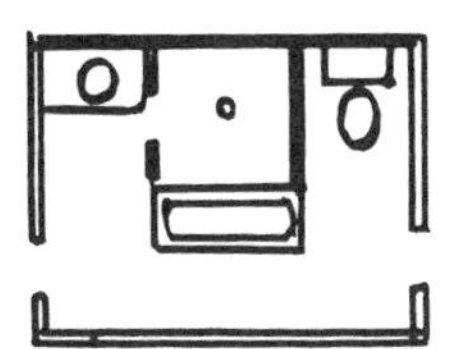

Fixture placement can provide privacy to different areas of the bathroom. Here, a shower stall divides the toilet and the lavatory. The room has two entrances.

The least expensive and easiest bathroom arrangement to install is one with all of the fixtures and plumbing pipes on one wall. Greater convenience and a better appearance may sometimes be achieved, however, by having fixtures on two or more walls. This may sometimes offset the additional one-time cost of piping.

A bathroom sink placed near the window will have good light. But it may be unwise to put the bathtub under a window because of drafts and because the window may be difficult to reach.

Plan enough floor space around fixtures for comfortable use and accessibility. For example, if a sink is placed next to a wall, allow 18 to 20 inches between the center of the sink and the wall for the arm movements required for shaving and grooming.

Fixtures

Since bathroom fixtures are the central focus in your bathroom plan, select their styles, sizes, colors, and materials carefully (see Chapter 7). But also pay attention to installing fixtures that suit the particular needs of your family.

For example, the normal installation height of a bath lavatory is about 33 to 36 inches from the floor. But if your family is made up of adults and teenagers, all of whom are built like basketball players, there is no reason why you can't place the sink 38 or 42 inches from the floor. On the other hand, if you have small children, you may wish to install a small second sink at a lower height. The same is true of shower heads. Normally they are placed about 74 inches from the floor. In a new installation, you can put it wherever it will do the most good for the most members of the family.

Storage

Storage space is essential to a well-planned bathroom. Provide plenty of storage space for bathroom linens, toiletries, cosmetics, and other items.

Cabinets for medicines and toilet articles usually are mounted above the bathroom sink. They have sliding or hinged doors (usually with mirrors), may have built-in lighting fixtures and electrical outlets, and come in

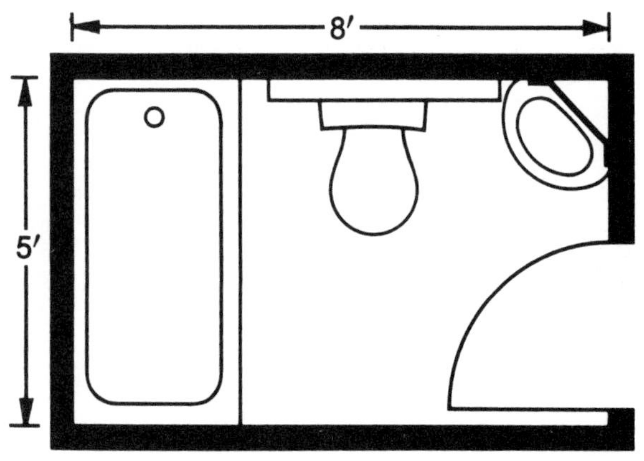

A western motif is featured in this small bathroom. The corner lavatory saves floor space, as does the storage cabinet built-in over the toilet.

many different sizes and styles. It is wise to have a separate cabinet or a compartment within a cabinet that can be locked. In this way medicines and other possibly harmful items can be kept away from children and half-asleep adults.

A convenient height for the medicine cabinet is 69 to 74 inches from the top of the cabinet to the floor.

Two shelves 12 inches deep and 18 inches wide, with 12 inches of clearance between them, are adequate for storing 12 bath towels and 12 washcloths. A hamper for soiled clothing is another convenience.

Space for storing linens and bathroom supplies should be included in even a small bathroom. Two likely storage places are under the lavatory and over the toilet tank. Leave at least 12 inches between the top of the tank and a shelf or cabinet hung above it to allow access to the tank when repairs are necessary.

Lighting

Adequate light, properly placed, is essential. For shaving and putting on makeup, light should shine on the face—not on the mirror. A ceiling fixture above the front edge of the sink and one light on each side of the mirror will illuminate the face without shadows. The side lights should be 30 inches apart.

Select light fixtures for bathrooms with white diffusers. Use white bulbs or shades (tinted ones distort colors). One fixture usually is adequate for a small bathroom. In large bathrooms, general illumination and area lights are needed.

One way to distribute light evenly is to install a luminous ceiling. These plastic-paneled ceilings can be installed in existing bathrooms.

Remember to install a grounded outlet at the bathroom sink. Make sure the height is convenient for electrical appliances.

Ventilation

Ventilation must be provided for all bathrooms. Often fans are used. Most automatic bathroom fans operate at only one speed, but some have as many as five.

For small bathrooms, an exhaust fan com-

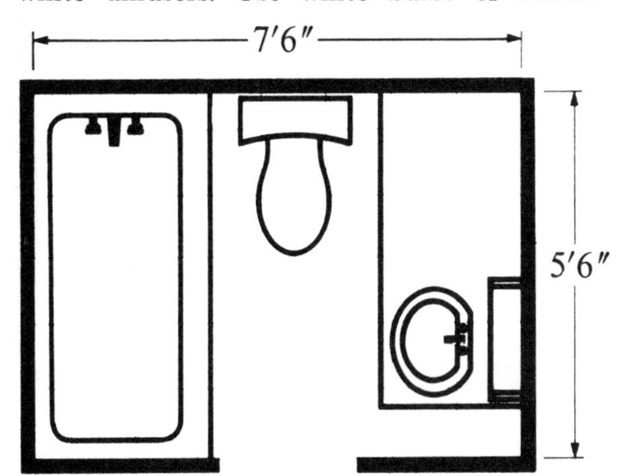

A spacious effect is achieved in a small bathroom by combining counter, medicine cabinet with banks of lights on each side, and a lighted ceiling as a single design element. Room colors, including fixtures, are bright and cheery. Dual shower heads serve both adults and children.

bined with a heater and lights is a good choice. They can be installed with one switch, but separate switches are preferable if such an installation is permitted by codes and ordinances.

You can get a dangerous shock while operating an electrical switch if you touch water and metal at the same time. Make certain that light and ventilating fan switches cannot be reached by anyone in the bathtub or shower or anyone using a water faucet.

Lights with pull cords are dangerous and should not be used in bathrooms.

Safety Measures

Bathrooms that will be used by the handicapped and the elderly need extra consideration. Safety, convenience, and ease of maintenance should not be forgotten. Doorways may have to be wide enough to accommodate a wheelchair or crutches. The door locks should be of the type that can be opened from either side.

Beware of slippery floors. Wall-to-wall carpeting with nonskid backing is preferable to scatter rugs. Install grab bars in tubs and shower stalls in places easy to reach whether one is sitting or standing. They must be mounted securely to the studs—a grab bar that pulls away from the wall under a person's weight is worse than none at all.

Grab bars on each side of the toilet make it easier for handicapped and elderly people to sit down and rise safely. Rubber mats with suction cups should be placed in the bottom of the tub to prevent slipping while sitting or standing. Night lights may be used for added safety.

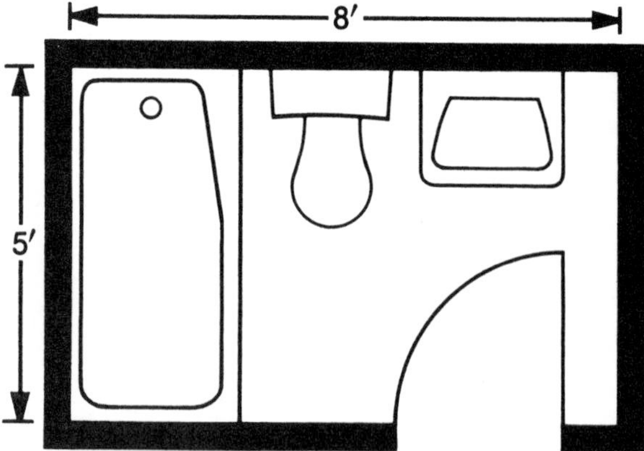

This sauna-style bath is another small room with a theme. The redwood walls and ceiling contrast with the gleaming white fixtures. The tapered pedestal of the lavatory adds to the Scandinavian look.

9 Plumbing Emergencies and Repairs

Once you understand your home's plumbing system, as well as its fixtures and fittings, its foibles will no longer seem so catastrophic. Even though you may not yet be fully acquainted with every ballcock, trap, and faucet spindle, you will be able to recognize their occasional malfunctioning for the nuisance that it is. You will be able to pinpoint the problem and, with the aid of your emergency toolbox, quickly get things working again.

While it may not be quite that simple, it isn't all that difficult either. For a system that does so much, the breakdowns are surprisingly few. And when things do go wrong, they usually can be put right again easily—now that you know what it's all about. For the most common problems, there are common solutions.

Leaky Faucets

A dripping faucet is more than just an annoyance. It is a water- (and therefore a money-) waster as well. Drips can stain the sink, and eventually they will wear away the finish too.

Before fixing a dripping faucet, shut off the water supply at the supply stop in the line below the faucet. (If there is no supply stop, turn off the water at some point further back in the line. And while you're at it, you may wish to install a supply stop, as described in Chapter 6.)

On some faucets, the handle covers the packing nut, and it must be removed to give access to the nut. On other faucets, the packing nut is exposed. Cover the nut with tape or rags to protect the chrome finish; then use an open-end or monkey wrench to remove the packing nut, turning it counterclockwise.

Remove the faucet stem or spindle and check the washer at its lower end. If the washer is worn or damaged, carefully remove the screw that holds it in place and take out the washer (if necessary, apply penetrating oil to loosen the screw). Replace the washer with a new one of the correct size. Use either a type-A or type-B washer (see illustration) if the rim around the stem bottom is intact. If the rim is damaged, file the remaining parts away completely and install a type-C washer.

1
When making any faucet repairs, first wrap tape around the packing nut to protect the faucet's surface.

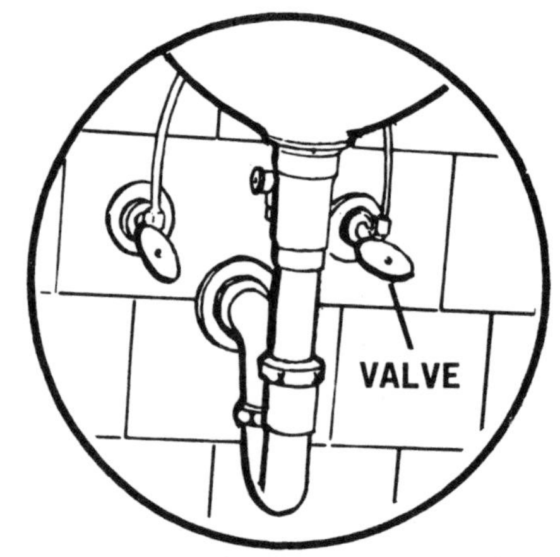

2
To repair a leaky faucet, first turn off the water at the supply stop valve.

3
Remove the packing nut, turning it counterclockwise.

4
Remove the stem or spindle from the body of the faucet.

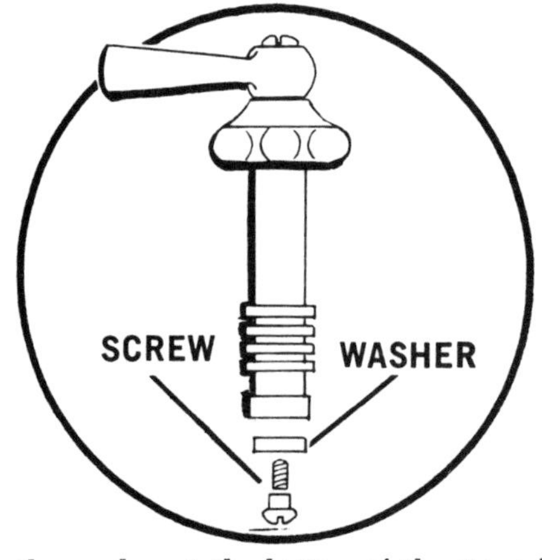

5
Check the washer at the bottom of the stem; if it is worn or damaged, remove the screw holding it in place.

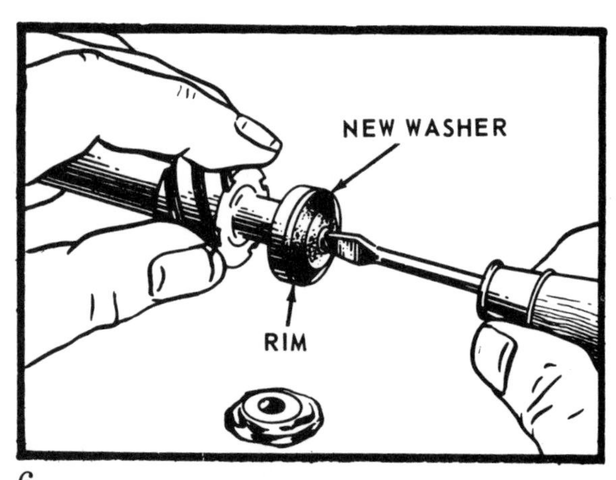

6
Replace with a new washer of the proper size; if the screw is corroded, use a new screw to hold the washer.

If the rim around the bottom of the stem is intact, use a type-A or type-B washer; if it is worn away, use a type-C washer.

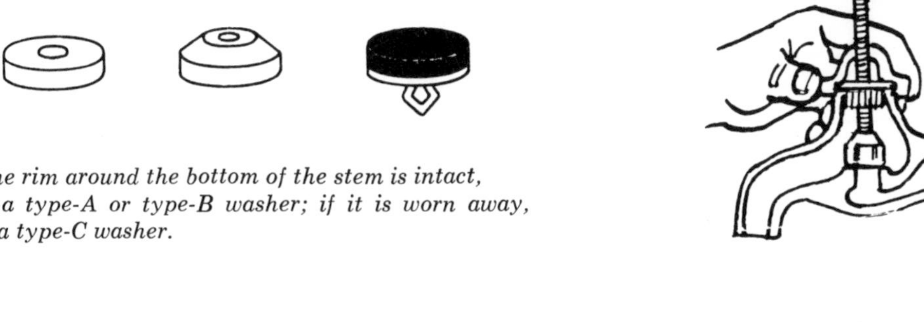

Hold the dressing tool in place and rotate it to grind the seat smooth.

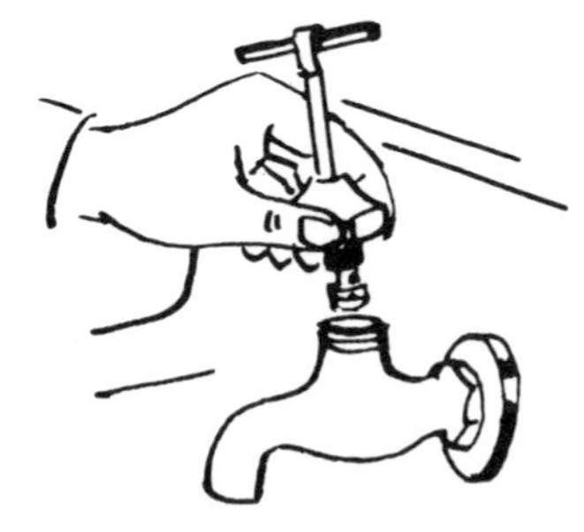

If the seat inside the faucet is nicked or damaged, repair it with a dressing tool.

On many faucets, a worn seat can be replaced. Remove it with an Allen wrench and install a new one of the same size.

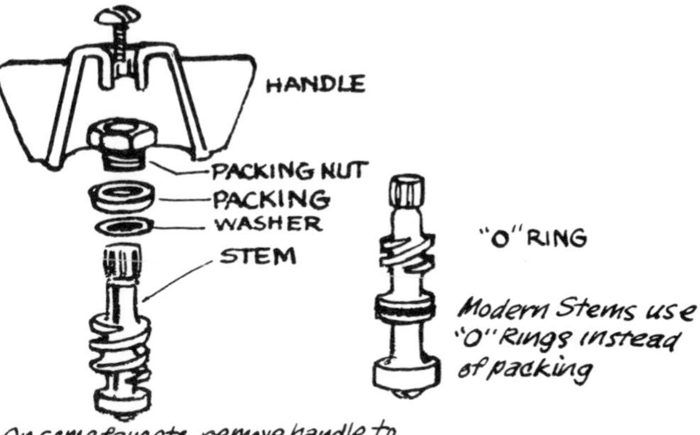

If a faucet continues to leak after the washer and seat have been repaired or replaced, examine the packing or O ring around the stem; replace it if necessary.

Shine a flashlight down inside the faucet body and make sure the seat is smooth and free of nicks. If not, it can be ground down with a seat-dressing tool. There are various types of inexpensive seat-dressing tools, but they all work in much the same way. The tool is clamped to the faucet (on some types it is held in place by the packing nut or bonnet) and the stem is turned by a T-handle; a fluted grinder at the other end of the stem reconditions the seat.

On many faucets, badly worn seats can be replaced. The old one is removed (usually with an Allen wrench) and a new one of the correct size is screwed into place.

After the seat is reconditioned and the old washer has been replaced, reassemble the faucet. If a leak continues around the stem, remove the handle and packing nut or bonnet; then examine the O-ring or packing around the stem. Replace if necessary.

On some types of tub faucets, the bonnet holding the stem in place is recessed in the wall. An extended socket wrench is needed to remove it for repair and for replacement of the faucet washer. Otherwise, the procedure is the same as described above.

Because the design and construction of non-compression (single-control mixer) faucets vary so greatly, it is all but impossible to set

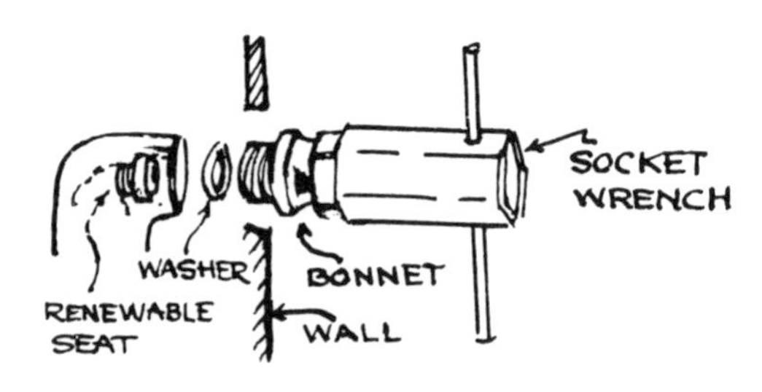

Some tub faucets are recessed, requiring a socket wrench to get at the parts for repairs.

general rules for their repair. Fortunately, most manufacturers include repair instructions as well as installation instructions. The best advice is to file away this repair sheet for future reference.

Most of the problems with these faucets occur when parts wear out. Replacement of the old parts is the only real remedy.

Noisy Faucets

A chattering faucet can be as annoying as a leaking one. First make sure that the faucet washer is tightly screwed to the stem. If the washer is worn, it should be replaced, even though the faucet may not be leaking. If the stem (after being screwed back into the faucet) can be moved up and down, it is an indication of worn threads and should be replaced. Occasionally, a design deficiency will cause faucets to chatter and whistle. A new faucet is the only remedy.

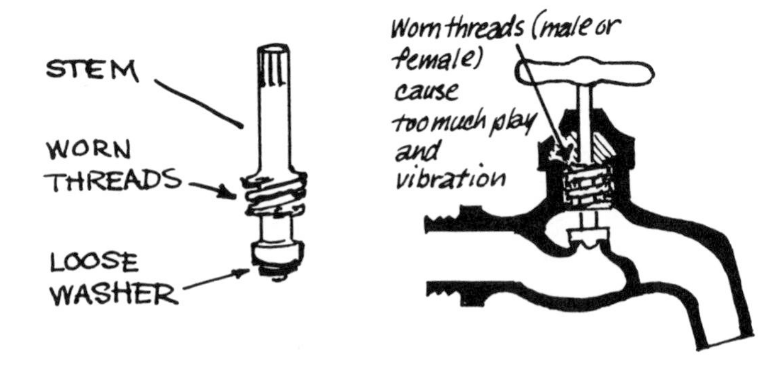

Chattering faucets may be caused by loose washers or worn stem threads.

Toilet Troubles

If the toilet tank overfills, the fault is usually with the ballcock. Check for a worn washer on the bottom of the plunger stem and replace it if necessary. Also check to make sure that the operating levers are moving freely. If they are frozen tight, you may have to replace the entire ballcock assembly (see Chapter 6).

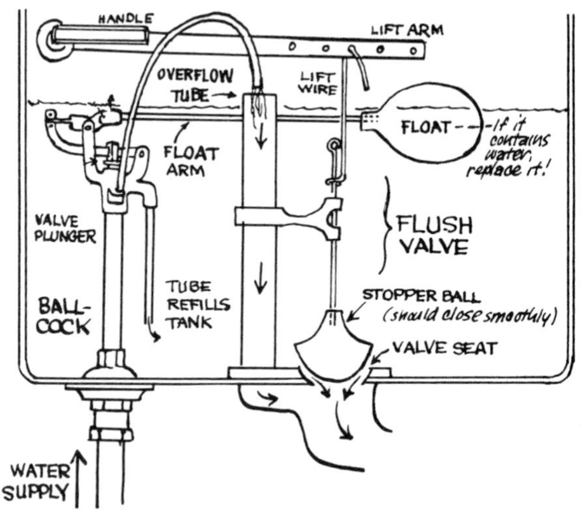

Toilet tank innards. Check out each of the components to see what is causing your problems.

Sometimes when a toilet is flushed, a fine spray of water will shoot up against the top of the tank and leak out around the cover. Check the overflow tube to see if it has a hole or if it is split. If so, replace it. The problem also could be a faulty or damaged connection between the overflow tube and the ballcock.

When the toilet will not stop running, with water going through the overflow tube, pull up on the float arm to see if this stops the flow. If it does, unscrew the float ball and check whether or not there is water inside. If there is, it has sprung a leak and should be replaced. If the float ball checks out, bend the float arm slightly to lower the ball. If this stops the water from running, all is well. If the flow still persists, turn off the water supply, remove the valve plunger, and replace the washer. If that doesn't do it, you can assume that the ballcock assembly should be replaced.

If the toilet continues to run because the tank does not refill, the problem is with the flush valve. Check the stopper guide to make sure it is aligned correctly. The stopper ball should drop squarely into the valve seat; adjust as necessary. If the stopper ball still does not seat properly, it may be damaged; inspect and replace it if necessary. Also check the valve seat, cleaning the surface with a fine emery cloth. If the valve seat is nicked or otherwise damaged, replace it.

If water keeps flowing into the tank, bending the float arm slightly to lower the ball may be the cure.

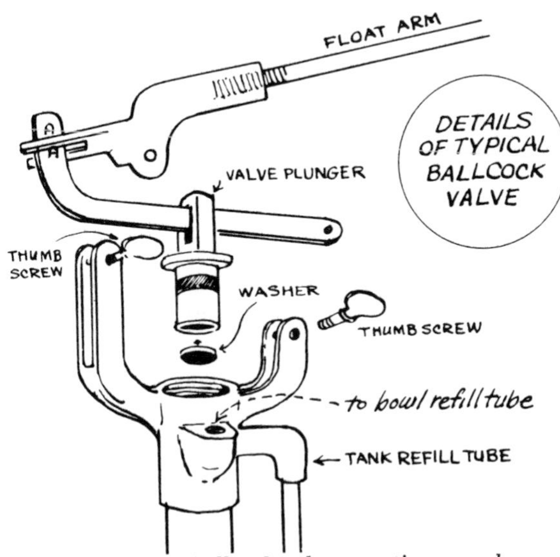

The washer in the ballcock valve sometimes needs replacement, or the moving parts may become stuck. If they are badly corroded, it is best to replace the whole mechanism.

Clogged Toilet

This can be a real mess, especially when the cruel fates conspire to combine it with another problem, such as a nonseating stopper ball which lets water pour continuously into—and out of—the toilet bowl. But never fear, a friend is near: the plumber's friend in your emergency toolbox. Grab that plunger and head for the trouble spot (you should have no trouble finding it). Place the suction cup over the discharge opening in the toilet bowl and work it vigorously up and down. This will usually clear the stoppage. Always make sure there is enough water in the bowl to cover the plunger cup while you are pumping.

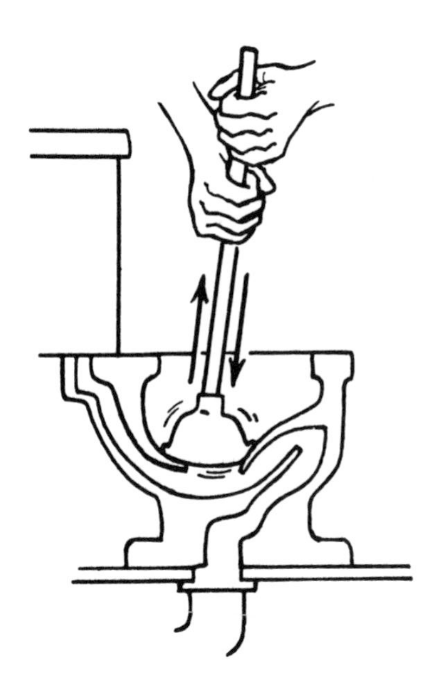

Your plumber's friend provides the easiest solution when a toilet clogs.

If your plumber's friend fails to do the job, it's time to bring on the snake, or auger. A closet auger is best for this job—it is designed to get the auger wire right down to where it's needed. But you can use a drain auger if that's all you have on hand. Take care not to scratch the inside of the toilet bowl.

Push the snake into the trap and crank it

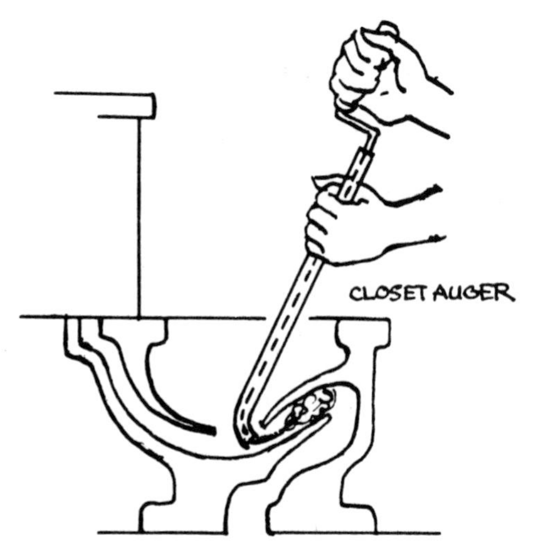

If the plunger will not work on an especially stubborn obstruction, use an auger to free the drain passage.

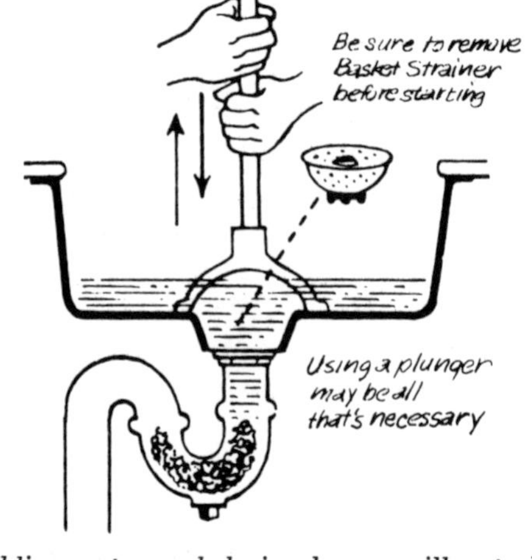

If scalding water and drain cleaner will not clear a clogged sink drain, try the plunger.

—in one direction only—until it becomes tight. Pull it back; often it will bring the obstruction up with it. If it doesn't, try again, cranking the snake until it pushes through and clears the trap.

As a last resort, remove the toilet from the floor (see Chapter 7), turn it upside down (resting it on old newspapers), and work the obstruction out through the discharge opening or, if it has already passed into the waste line and is stuck there, work it out with the snake.

When replacing the toilet, follow the instructions in Chapter 7 for setting a new toilet. Use a new wax gasket around the opening and apply fresh putty around the rim.

Clogged Kitchen Sink Drains

This is another annoyance that people often bring on themselves by removing the strainer and letting food wastes or grease run into the sink drain.

If the drain is not completely clogged, you may be able to clear it by letting scalding water run into it for several 'minutes. If that doesn't work, try a dry or liquid chemical drain cleaner (many contain lye, but *do not use lye by itself*). Follow the manufacturer's instructions carefully. If the chemical clears the drain, flush it with hot water for at least 10 minutes.

You may be able to remove the obstruction through the cleanout plug in the sink trap.

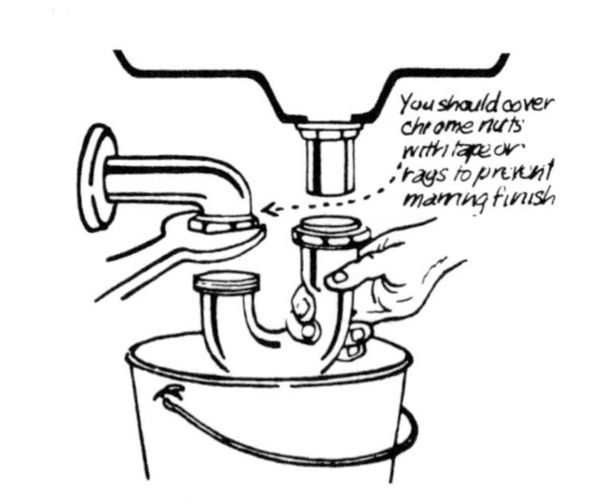

The next move is to remove the trap, but first place a bucket beneath the sink to catch the water and debris.

When all else fails, use a snake.

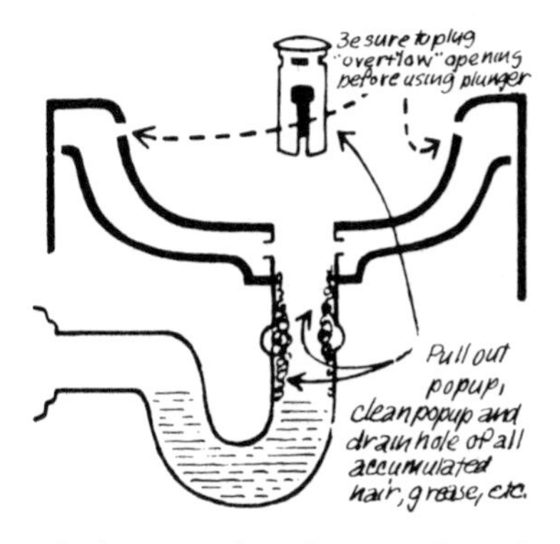

Lavatory drains sometimes become clogged when hair and other waste get caught in the pop-up plug.

If the chemical doesn't do the job or if the drain is completely clogged, call on your plumber's friend. Remove the strainer from the drain and make sure there is enough water in the sink to cover the plunger and provide a seal. Work the plunger up and down—and hope for the best. This procedure will usually work.

If your friend fails, check the trap below the sink. If it has a cleanout plug, remove it—you may be able to get at and clear the obstruction. If you still can't clear the drain, or if there is no cleanout plug, loosen the slip nuts and remove the trap. If something is caught in the trap, just take it out and your troubles are over. If not, use the drain auger. Feed it into the drain line and rotate it; then feed it in some more and rotate it again. Repeat this procedure until the obstruction is cleared.

Reassemble the trap, then run scalding water down the drain for several minutes to carry away grease and other accumulations.

Clogged Lavatories

Twist the pop-up plug to disengage it and lift it out of the drain. Often the sink has been clogged by an accumulation of hair, grease, and the like on the plug. Clean it off. With the plug removed, you also can use a length of wire with one end bent into a hook to fish out any debris that may be stuck between the discharge opening and the trap.

If these efforts don't solve the problem, proceed with the measures described above under Clogged Kitchen Sink Drains. Plug up the overflow outlet with rags before using the plumber's friend; otherwise it cannot do its job properly.

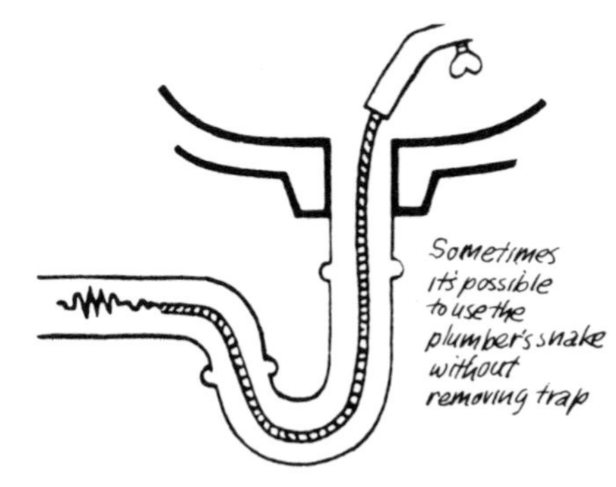

On some drains, you may be able to use the auger without removing the trap. Be careful not to scratch the surface of the basin.

Clogged Bathtubs

Try scalding water, then chemicals, then the plunger, as with sink drains. If these don't work, you will have to remove the trap.

Some first-floor bathtubs have P- or U-traps that are accessible from the basement or crawl space. In this case, loosen the nuts, take off the trap, and clear the line with an auger, as described above.

Other tubs have drum traps, usually located near the bathtub drain, with an access cover flush with the floor. Make sure the tub is empty before opening the drum trap—otherwise you will have a geyser at the trap.

With an open-end wrench (never use a pipe wrench) remove the drum trap access cover

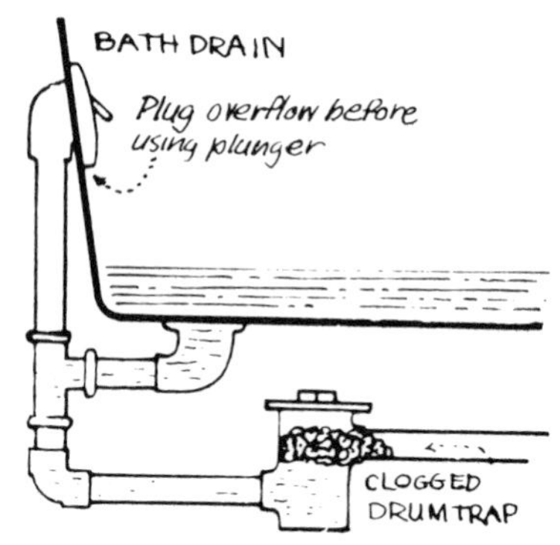

When other methods fail to clear a clogged bathtub drain, attack the problem through the drum trap.

and the rubber gasket. Clean out the trap. Work the drain auger in the line between the tub and the trap. If the obstruction is not found there, work the auger in the outflow line from the trap until the line is cleared. Check the gasket and install a new one if necessary; then replace the cover on the trap. Run hot water into the drain for several minutes to be certain the obstruction has been cleared.

Clogged Sewer

When drainage is sluggish (or completely lacking) from several or all fixtures in the house, the obstruction is in the main drain. Go to the basement cleanout plug. Place a bucket below the plug and use a wrench to loosen the plug just enough to allow the water to flow out into the bucket. When all of the water has drained out, remove the plug.

Remove the nozzle from your garden hose and push the hose into the cleanout opening until it reaches the obstruction. Wrap damp rags around the hose at the opening until it is tightly sealed. Have another person slowly turn on the water as you force the hose tightly against the stoppage. As the obstruction begins to give way, increase the water pressure. If there is no give, turn off the water and remove the hose.

Next try a drain auger. Turn it into the drain until it hits the obstruction; then push and pull while turning the handle of the auger until the obstruction is cleared away. Then run the hose into the drain to wash down the debris.

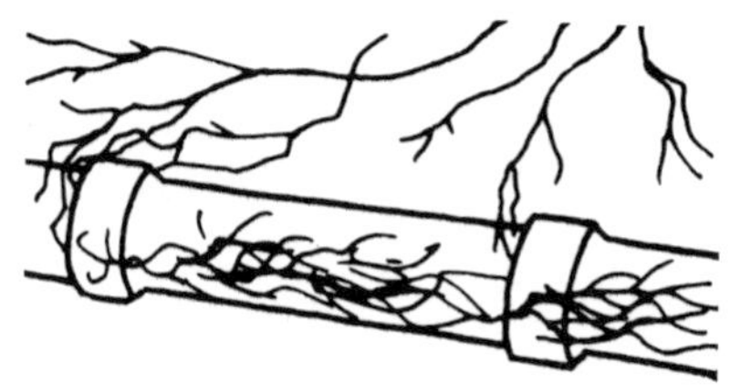

Clogged main drains frequently are caused by tree roots penetrating faulty joints in the line between the house and the sewer or septic tank.

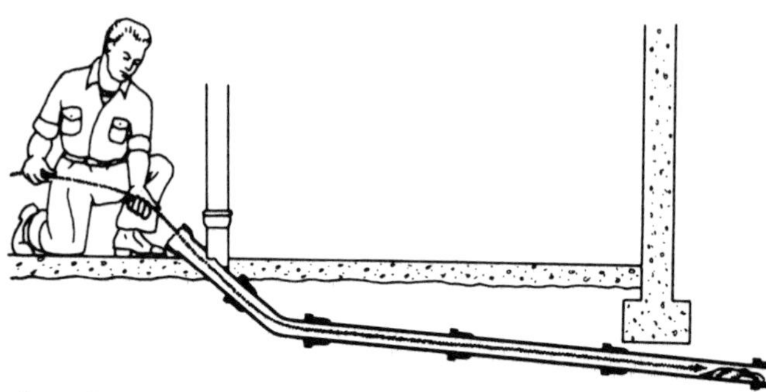

As a last resort, an electric auger can be rented to clear obstructions in the building drain.

If the problem persists, it may be caused by tree roots that have penetrated the sewer line. Electric augers with sharp blades for cutting through roots can be rented. Make sure you follow the dealer's instructions when using such a machine. After the obstruction has been cleared, use the hose to rinse out the drain before replacing the cleanout plug.

Leaking Pipes

When a pipe springs a leak, turn off the water immediately. You can make a temporary

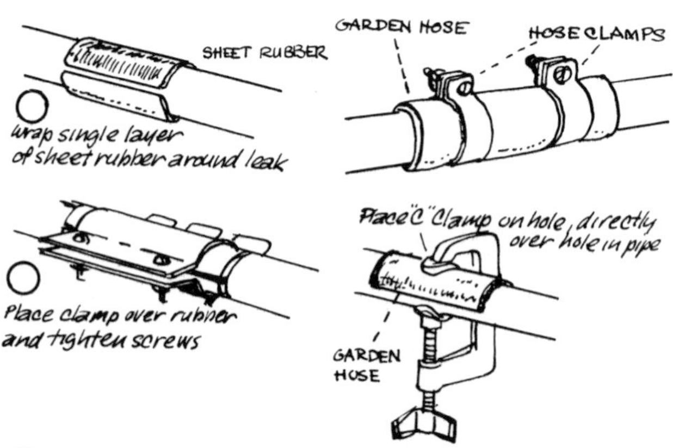

Emergency repairs for a leaking pipe.

repair by wrapping sheet rubber over the leaky section of pipe and clamping it tightly. Or, you can cut a section of garden hose, slit it, place it over the leak, and hold it there with a C-clamp. If the pipe is actually broken, the two sections can be rejoined with a short piece of garden hose, with hose clamps on each end.

To make a permanent repair, cut out the damaged section with a cutter or hacksaw. Fit a coupling to one of the remaining sections, cut a short piece of pipe to fit, and attach it to the other section with a union (see Chapters 3 and 4).

If the leaking pipe is within a wall, you have to break through the wall to make the repair. Therefore, it is easier to disconnect the pipe from the supply line and the fixture it serves and to run a new line next to the old one. Soft copper tubing is best for snaking through walls.

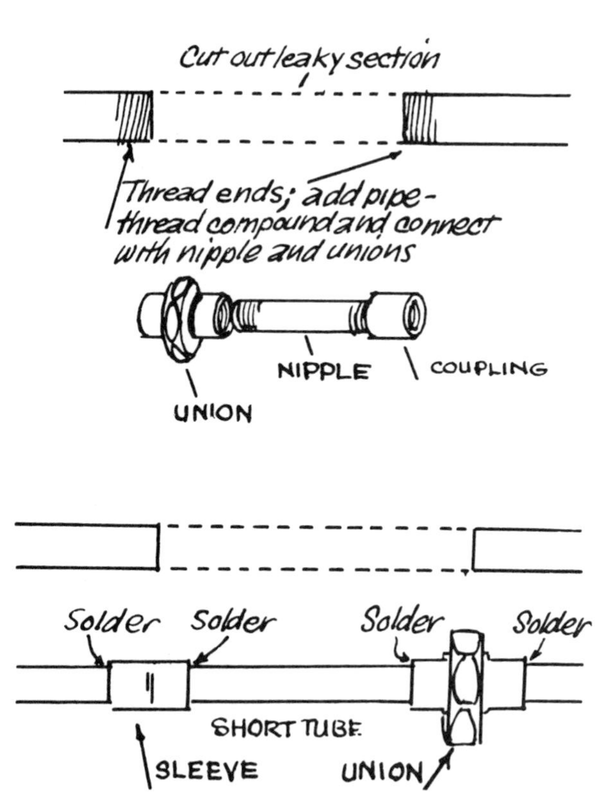

A damaged pipe is permanently repaired by cutting out the damaged section and replacing it with new pipe. A union fitting simplifies the job.

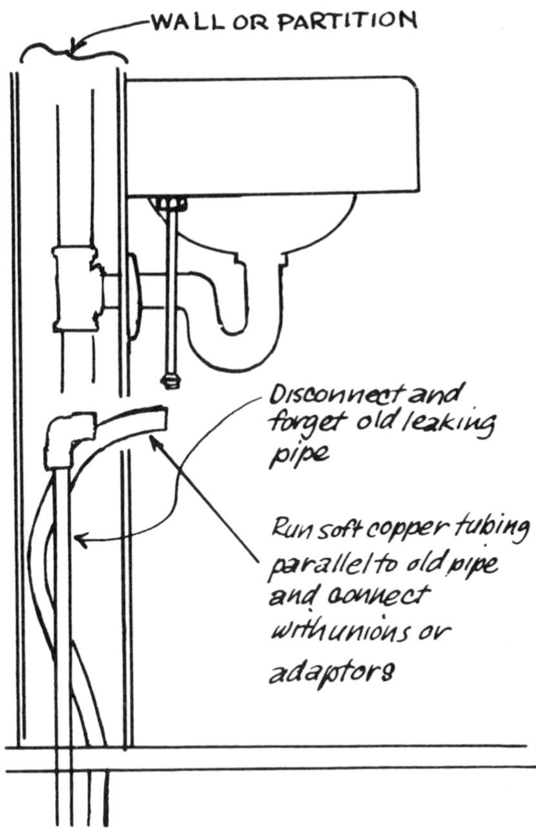

When a pipe within the wall springs a leak, it is best to forget it and to install a new pipe parallel to the old one.

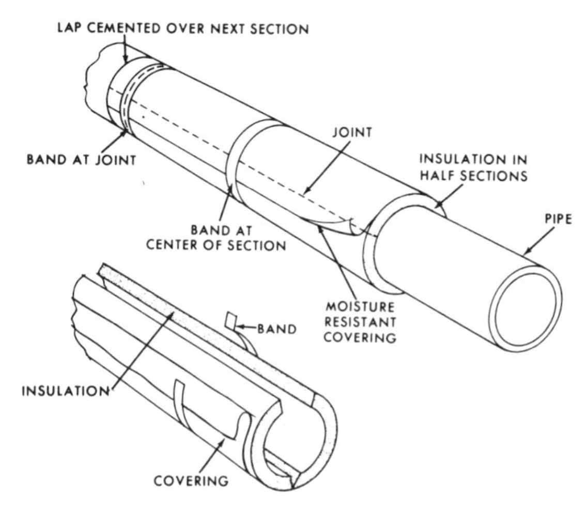

"Sweating" pipes can be wrapped with insulation.

Sweating Pipes

Pipes that "sweat" during hot weather are an aggravation, and they are more than that if you plan to finish off your basement—they can cause damage to the ceiling. The cure for sweating pipes is simple: wrap them with insulation. Insulation doesn't just catch the drips—it stops them from forming by helping the pipe keep its cool. You can use asbestos or fiberglass insulation, or you can use a special thick tape made for this purpose.

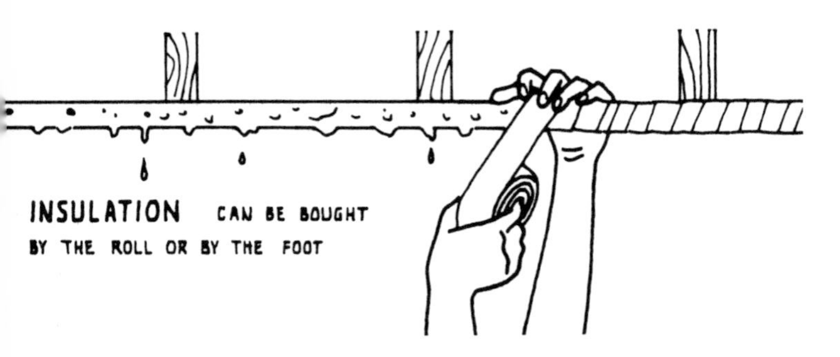

A special tape also is made for wrapping pipes to prevent condensation from forming.

Frozen Pipes

When water pipes freeze they may burst, or joints may be forced open. They should be thawed quickly but with caution.

First, open the outlets in the line. The safest way to raise the temperature is to wrap cloth or burlap around the frozen pipe and to pour boiling water over the cloth. Keep it saturated with hot water until the water in the lines is flowing freely.

The safest way to thaw frozen pipes is to wrap them with rags or burlap and to pour on boiling water.

A blowtorch or propane torch should be used only if the pipe is in a safe location—away from flammable walls or other materials—and then it should be moved along the pipe rather than concentrated on a single spot. Start near an open outlet and work toward it (never away from it), so that steam generated within the pipe can escape.

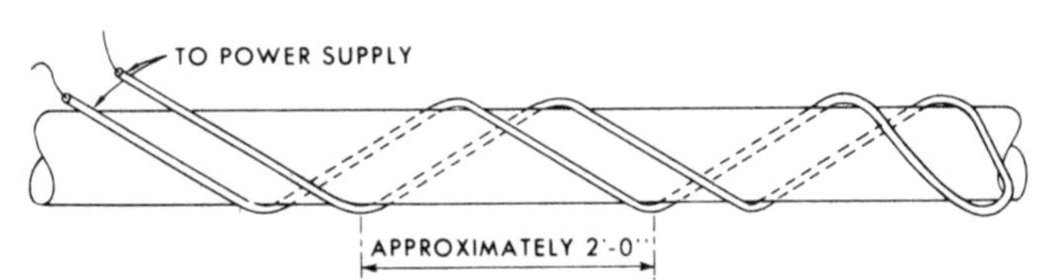

An electric heating cord can be used to thaw frozen pipes. It is also a good preventive measure for pipes in cold areas of the basement or crawl space.

10
Septic Tanks and Cesspools

If you live in a community that enjoys the benefits of municipal sewage disposal, consider yourself fortunate. Many homeowners in suburbia and ruralia are not so fortunate and have to depend on a septic system for final waste disposal. While this is certainly an improvement over the privy and other even more primitive disposal systems, the septic set-up is nevertheless problematic and requires some care and monitoring to prevent your awakening some malodorous morning to find your backyard awash.

How It Works

The typical septic system consists of a sewer line that carries wastes from your house, a tank into which it discharges, and an absorption field where liquids drain away into the ground. If you have an older disposal system, or are located in a really remote area where local health ordinances have not yet penetrated, you may have a cesspool instead of a septic tank. A cesspool does not include an absorption bed. It usually consists of an open-bottomed tank, or one with perforated sides, or even just a hole lined with rocks, into which wastes are dumped from the sewer drain. Your nose knows which system you have.

Both systems are supposed to trap solids while permitting liquids to drain away into the surrounding soil. The tank is where the action is. Its purpose is to liquefy solid matter contained in the sewage collected in it. This is accomplished by the dissolving action of bacteria in the waste. This chemical reaction results in a clear liquid which flows from the tank to the disposal field. In a septic system, the disposal field is an absorption bed of perforated pipe through which the outflow seeps into the surrounding soil for final disposal. In a cesspool, the liquid is supposed to seep out the sides or through the bottom of the tank or hole.

The septic tank is usually made of reinforced concrete. Inside the tank, near the inlet and outlet, baffle-plates direct the waste flow for greatest efficiency. The solids settle to the bottom, where bacterial action separates them into a sludge, which remains on the bottom of the tank, and a scum, which rises to the top. The layer of scum is supposed to break down gradually into

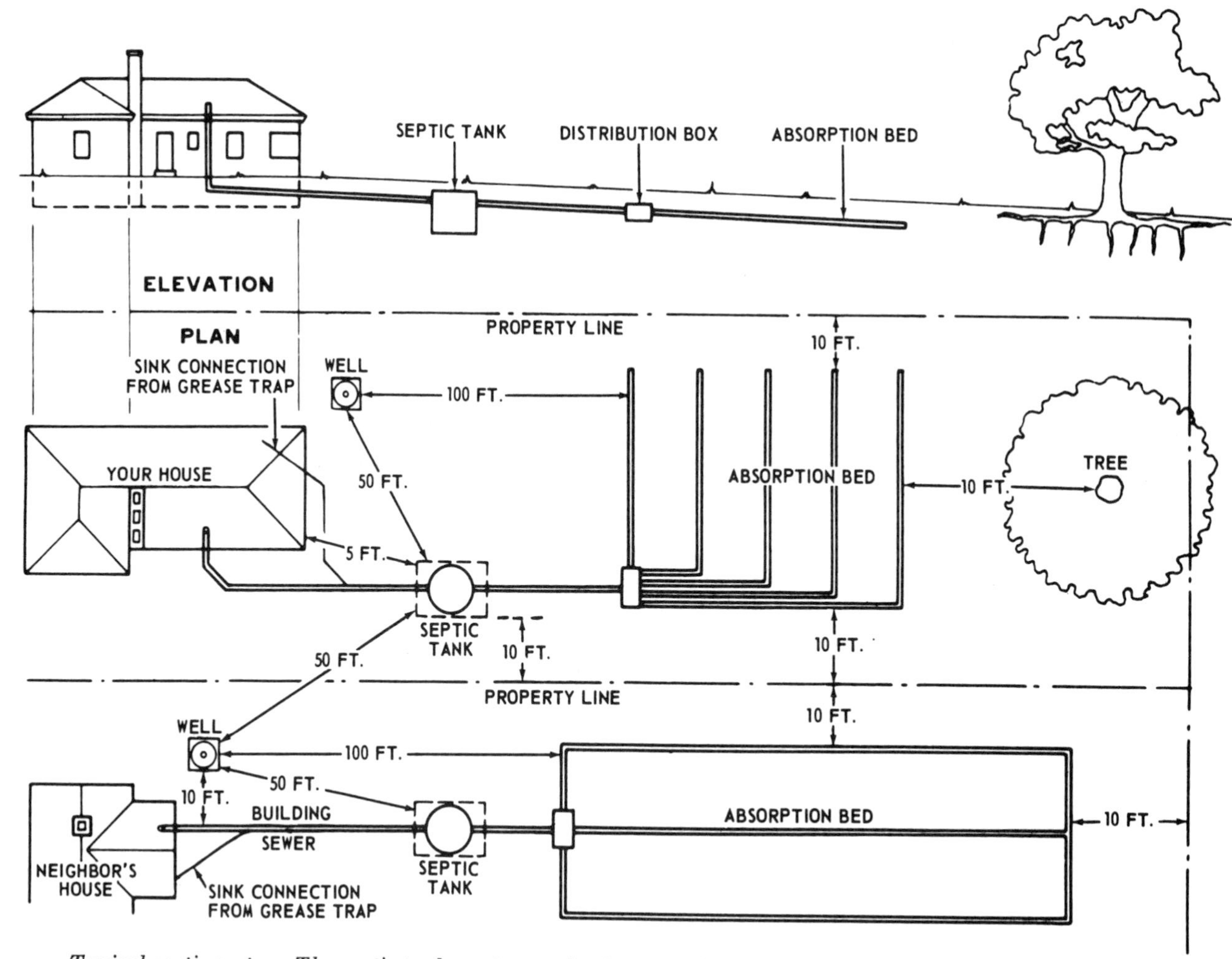

Typical septic system. The septic tank must never be situated within 50 feet of the water supply, and the absorption bed must be at least 100 feet from a well. Most building codes also include restrictions on how close your septic system can be to that of your neighbor.

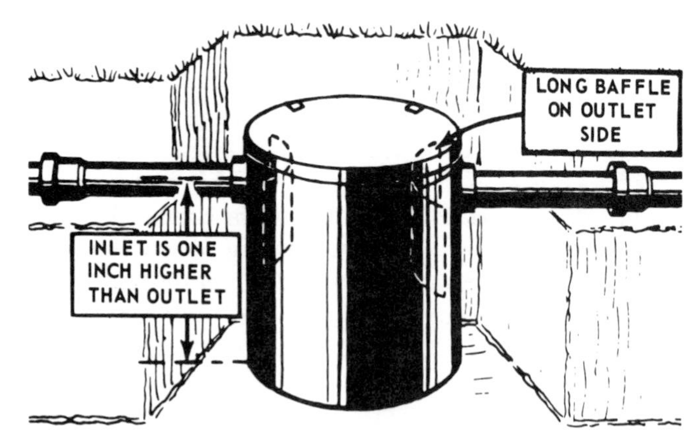

Septic tanks may be round, square, or rectangular. Minimum capacities are usually set by local health departments (see accompanying table). Baffles inside the tank control flow of wastes.

liquid in the middle of the tank, and ultimately run out through the outflow line into the absorption bed.

For most efficient operation, the first stop along the drain line from the septic tank should be a distribution box, which regulates and equalizes the flow in all lines of the absorption bed. The distribution box, like the rest of the system, is buried out of sight, but is usually close enough to the surface to serve as an inspection manhole, allowing you to check the quality of the effluent (liquid sewage after it leaves the septic tank) and to see if sludge particles are being carried out into the absorption bed. If so, it would mean trouble.

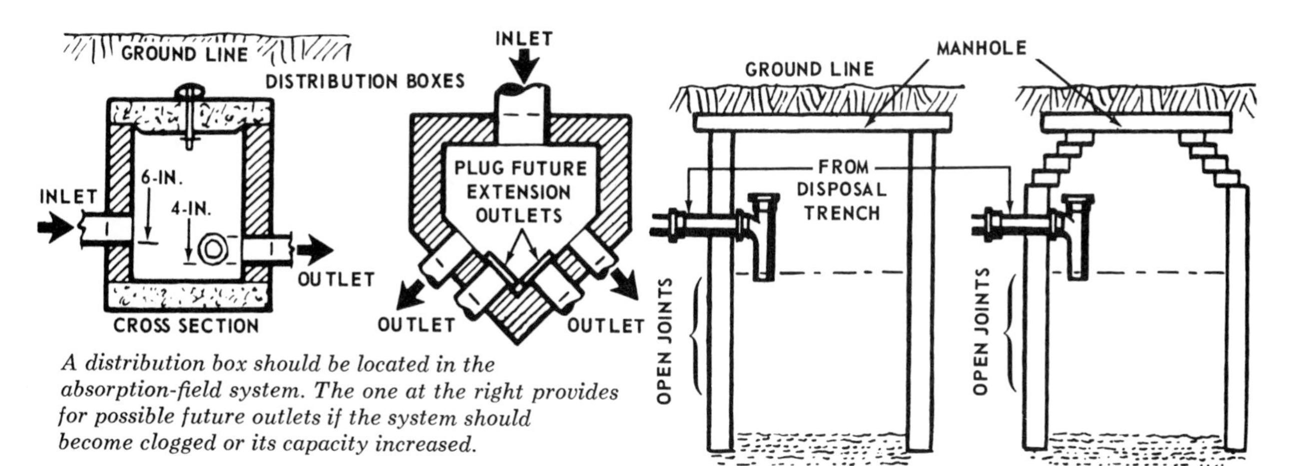

A distribution box should be located in the absorption-field system. The one at the right provides for possible future outlets if the system should become clogged or its capacity increased.

To keep rainwater and foundation water away from the septic tank, a seepage pit may be provided. Cement pipe with open (uncemented) joints, laid on a gravel bed, will serve the purpose.

Problems

A balance must be maintained among the three tank ingredients: sludge, liquid, scum. If heavy layers of sludge build up and overflow into the absorption bed, they can clog the disposal lines and the whole system, with the messy result that raw sewage will be forced to the surface. If it comes to that, all you can do is dig up the yard and install a new system. There is virtually no way to unclog the old one.

One important preventive device is a grease trap, either at the kitchen sink or immediately outside the house near the sink. Its purpose is to keep grease, fats, detergents and other kitchen wastes (except for liquids) out of the septic tank, where they contribute to too-rapid buildup of solids in the tank bottom and (as is especially true in the case of detergents) refuse to break down by the bacterial action in the tank.

Rainwater, surface water, and water from around the house foundation must never be allowed to drain into the septic tank, upsetting the balance. If necessary, a seepage pit should be provided to catch drainage from roofs, areaways, and basement floors. This should be located at least 20 feet away from the septic system absorption field. The seepage pit can be simply a 3-foot length of large-diameter vitrified clay- or cement-pipe covered with a manhole and buried in the ground on a bed of coarse gravel or crushed stone that will permit water to seep into the soil. Drainpipes from gutters and other water-collectors empty into the seepage pit.

Most septic-tank households quickly develop conscientious water-use habits, especially after they have experienced an overflow. Leaky faucets and running toilets loom as major catastrophes, and demand immediate repair (Chapter 9). Baths are drawn less deep, toilets flushed less frequently. The laundry, a large consumer of water, also poses a problem. Sometimes, a completely separate septic system is provided for the laundry facilities, where solid wastes are less likely to build up.

Normal Maintenance

Eternal vigilance is the price of a smooth-running septic system. Even with vigilance things can go wrong, but at least you know that the problems are not your fault.

If you are building a new house or having a new septic system installed for an existing house, make a map of the system, showing the location of the tank and the absorption field, including the distribution box. This will be an invaluable aid when it comes to pinpointing problems.

Cleaning of the tank must be done periodically, and is a job best left to the professionals with their heavy equipment. How often this must be done will depend on several factors: the number of persons served by the system, daily sewage flow, porosity of the soil, capacity of the tank. With ordinary use, cleaning of the average septic tank should be necessary every two to three years.

Inspection should be much more frequent than that—every 12 months or so. In most cases, this means digging down to the tank opening to check the depth of accumulated sludge. If you have a service contract with a septic-tank pumping firm, they will probably do this for you at regular intervals; just make sure they do. A special measuring stick is used for the job, and the general rule is that when the scum accumulations and the sludge layer reach a combined depth one-third that of the liquid in the middle, the tank should be cleaned. Don't put it off.

It is best to plan ahead, so that sludge is removed in the spring rather than in the fall. This will avoid loading the tank with undigested solids during the cold winter months, when bacterial action slows down considerably. If you elect to do the cleaning job yourself, sludge disposal must be by burial or other method approved by your local health authorities. The sludge may include disease-carrying bacteria, another reason why this job is best left to the professionals.

Recommended Septic Tank Capacities *

Number of bedrooms in house	Capacity per bedroom in gallons	Required total tank capacity in gallons
2 or less	375	750
3	300	900
4	250	1000
5	250	1250

For more than 5 bedrooms, add 250 gallons per bedroom.

* U.S. Public Health Service Recommendations. Check local health authorities for regulations in your area.

Capacities given above provide for use of garbage disposals, automatic washers, and other household appliances. Some regulations permit sizes 150 gallons smaller if garbage disposals are not used.

11 Preventive Maintenance

All things considered, your home's plumbing system spends a long life doing a tremendous amount of work with a minimum of trouble. You can help it to live even longer and do its jobs better by faithfully practicing a few "ounces of prevention."

Make sure all members of the family know where the main water shutoff valve is located, just in case an emergency arises when the home's master plumber is not on the premises. It's a good idea, too, to tie tags on basement valves and water lines, labeling them hot and cold and telling which fixtures they supply. This is another help when a quick shutoff is necessary.

Care of Fixtures and Fittings

Protect porcelain enamel surfaces of sinks, lavatories, and bathtubs by washing them frequently with hot water and soap. If you use a cleanser, make sure that it is the nonabrasive type.

Some kitchen sinks are specially treated to resist staining; but lemon, orange, and other citrus fruit juices, tomato juices, mayonnaise and other vinegar preparations, and tea and coffee grounds should not be allowed to remain long on the enamel surfaces; wash them away as soon as possible. Be careful not to scrape pots and pans across the sink basin or drainboard, and never use the sink or drainboard as a cutting board or chopping block.

Keep grease, coffee grounds, and food particles from entering the sink drain. A certain amount of grease will probably go down the drain when you rinse pans and dishes; this grease deposit will trap coffee grounds if they are allowed into the drain, gradually building up to a stoppage. Use drain cleaners on a regular schedule to prevent stoppages; they will save you the aggravation of more elaborate drain-clearing operations later on.

Treat the bathroom lavatory to a regular diet of drain cleaner as well. Frequently lift out the pop-up drain plug and clean it thoroughly. Keep articles in the medicine cabinet neatly arranged so that glasses, bottles, jars, and the like don't fall accidentally into the lavatory. If acid or medicine should spill on the lavatory surface, wash it off immediately.

The vitreous china of toilet bowls and tanks is impervious to most acids.

Familiarize all members of the family with the location of the main shutoff valve and with its operation.

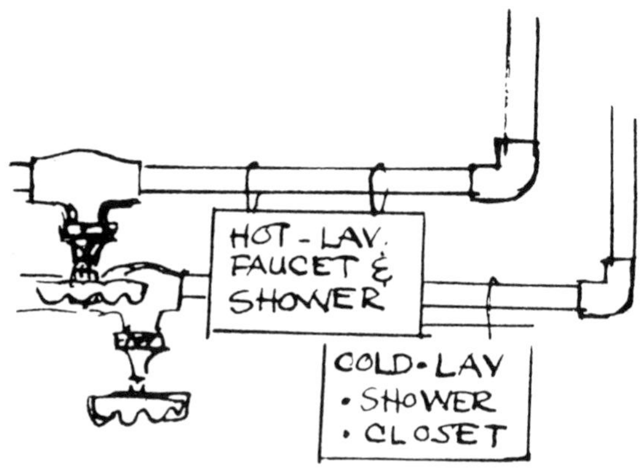

It's a good idea to put labels on all exposed pipes in the basement, telling what each is and where it goes.

Clean porcelain enamel surfaces and chromium-plated fittings with hot water and mild soap. Never use abrasive cleansers.

Regularly pour drain cleaner into the kitchen sink and bathroom lavatory drains to prevent stoppages.

Lift off and clean-out the pop-up drain plug in the lavatory.

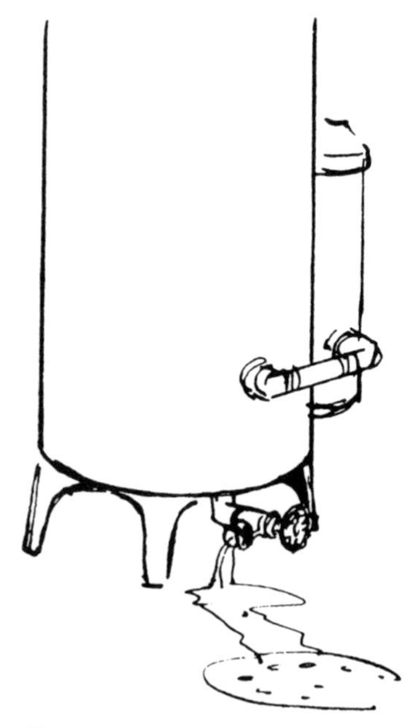

To prevent sediment buildup in the water heater tank, drain water from the bottom faucet at regular intervals, until it runs clear.

Hot water and soap should be enough to clean the toilet, but if something more is needed, use a nonabrasive cleanser.

Chromium-plated fittings require little attention other than an occasional washing with mild soap and water or a specially prepared liquid cleaner. If the fittings have been damaged, green spots may appear, indicating corrosion. To prevent the corrosion from spreading, scour the spots with a powdered cleanser until they disappear. Then apply a film of wax to the fitting.

Water Heaters

Keep the temperature setting of your home's water heater at or below 140° F. That should be hot enough for any household needs and will not only save fuel but also prevent excessive wear on the heater.

To prevent sediment from accumulating in the heater tank, drain a few quarts of water out of the bottom several times each year. Do the job in the morning, when the sediment has settled. If the heater is gas-fired, shut off the gas valve. Close the water valve at the heater and open a faucet to prevent a vacuum. Drain water from the valve or faucet at the bottom of the tank until the water runs clear.

There should be a relief valve at the top of the water heater, a safety feature to prevent damage from excessive pressure and temperature. This valve may become stuck or corroded from long disuse. When you drain the tank, check the valve by tripping the lever manually to make sure it will operate freely if an emergency arises.

Shutting Down the System

If your home or vacation house is going to stand vacant during the cold months with the heating system off, the plumbing system must be closed down.

First shut off the main water supply valve. Starting on the top floor, open all faucets in the house and leave them open. When water has stopped running from all faucets, open the plug in the main water supply valve and drain the remaining water into a pail.

Drain the hot water tank, making sure it completely empties. Make sure all horizontal pipes in the system are drained. If necessary, disconnect pipes to drain them.

Remove water from the traps under sinks, lavatories, and tubs by opening cleanout plugs in the traps (if they have them) or by removing the traps and pouring out the water (see Chapter 6). Empty the toilet bowl and tank as described in Chapter 6. Sponge out any water that remains in the toilet tank. Add antifreeze to the toilet bowl (about two quarts) and to all other fixture traps, as well as to the trap in the basement floor drain.

When a plumbing system is shut down in the winter (as in a vacation home), drain all water from supply lines, heater, toilets, and traps. Pour antifreeze into toilets and traps.

Index